HACKING ELECTRONICS

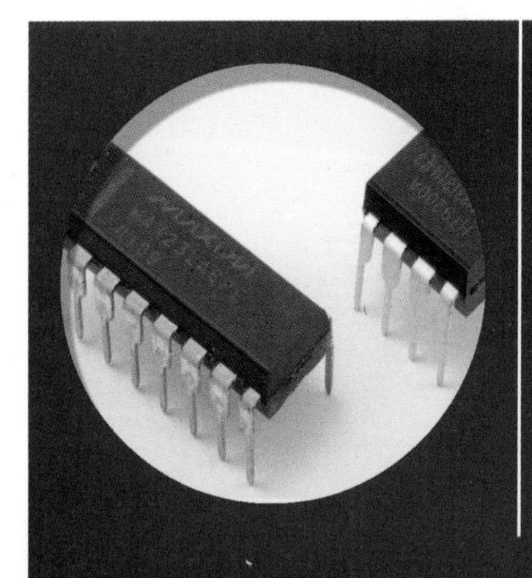

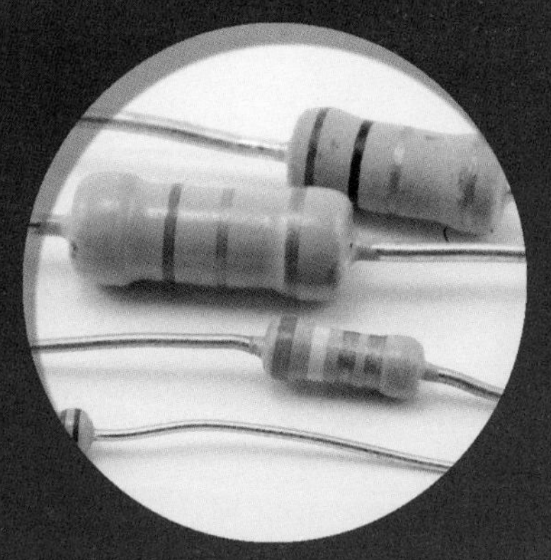

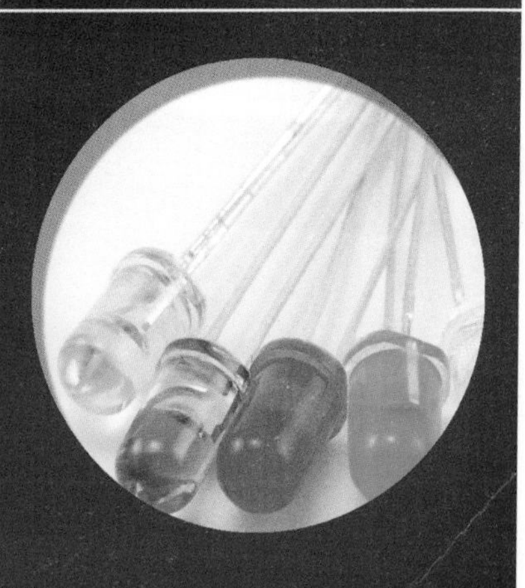

About the Author

Dr. Simon Monk (Preston, UK) has a degree in Cybernetics and Computer Science and a PhD in Software Engineering. Monk spent several years as an academic before he returned to industry, co-founding the mobile software company Momote Ltd. He has been an active electronics hobbyist since his early teens and is a full-time writer on hobby electronics and open-source hardware. Dr. Monk is the author of numerous electronics books, specializing in open-source hardware platforms, especially Arduino and Raspberry Pi. He is also co-author with Paul Scherz of *Practical Electronics for Inventors,* 3rd edition. You can follow Simon on Twitter, where he is @simonmonk2.

Hacking Electronics

An Illustrated DIY Guide for Makers and Hobbyists

Simon Monk

New York Chicago San Francisco Lisbon
London Madrid Mexico City Milan New Delhi
San Juan Seoul Singapore Sydney Toronto

Cataloging-in-Publication Data is on file with the Library of Congress

McGraw-Hill books are available at special quantity discounts to use as premiums and sales promotions, or for use in corporate training programs. To contact a representative, please e-mail us at bulksales@mcgraw-hill.com.

Hacking Electronics: An Illustrated DIY Guide for Makers and Hobbyists

5 6 7 8 9 10 DOC 21 20 19 18 17 16

ISBN 978-0-07-180236-9
MHID 0-07-180236-3

Sponsoring Editor Roger Stewart
Editorial Supervisor Jody McKenzie
Project Manager Vastavikta Sharma, Cenveo® Publisher Services
Acquisitions Coordinator Amy Stonebraker
Copy Editor Mike McGee
Proofreader Claire Splan
Indexer James Minkin
Production Supervisor James Kussow
Composition Cenveo Publisher Services
Illustration Cenveo Publisher Services
Art Director, Cover Jeff Weeks

To Roger, for making it possible for me to turn a hobby into an occupation.

Contents at a Glance

Contents

Acknowledgments

Many thanks to all those at McGraw-Hill Education who have done such a great job in producing this book. In particular, thanks to my editor Roger Stewart and to Vastavikta Sharma, Jody McKenzie, Mike McGee, and Claire Splan.

Special thanks are due to Duncan Amos, John Heath, and John Hutchinson for their technical review of the material and encouragement.

And last but not least, thanks once again to Linda, for her patience and generosity in giving me space to do this.

Introduction

This is a book about "hacking" electronics. It is not a formal, theory-based book about electronics. Its sole aim is to equip the reader with the skills he or she needs to use electronics to make something, whether it's starting from scratch, connecting together modules, or adapting existing electronic devices for some new use.

You will learn how to experiment and get your ideas into some kind of order, so that what you make will work. Along the way, you'll gain an appreciation for why things work and the limits of what they can do, and learn how to make prototypes on solderless breadboard, how to solder components directly to each other, and how to use stripboard.

You will also learn how to use the popular Arduino microcontroller board, which has become one of the most important tools available to the electronics hacker. There are over 20 examples of how to use an Arduino with electronics in this book.

Electronics has changed. This is a modern book that avoids theory you will likely never use and instead concentrates on how you can build things using readymade modules when they are available. There is, after all, no point in reinventing the wheel.

Some of the things explained and described in the book include

- Using LEDs, including high-power Lumileds
- Using LiPo battery packs and buck-boost power supply modules
- Using sensors to measure light, temperature, vibration, acceleration, sound level, and color
- Interfacing with Arduino microcontroller boards, including using Arduino shields such as the Ethernet and LCD display shields
- Using servo and stepper motors

Some of the things described in the book that you can make along the way include

- A noxious gas detector
- An Internet-controlled hacked electric toy
- A device for measuring color
- An ultrasonic rangefinder
- A remote control robotic rover
- An accelerometer-based version of the "egg and spoon" race
- A one-watt audio amplifier
- A bug made from a hacked MP3 FM transmitter
- Working brakes and head lights that can be added to a slot car

You Will Need

This is a very practical, hands-on type of book. You will therefore need some tools and components to get the most out of it.

As far as tools go, you will need little more than a multimeter and soldering equipment.

When it comes to areas of electronics where a microcontroller would be useful, an Arduino Uno board is best. So you may wish to buy one of these microcontroller boards before attempting some of the projects.

Every component used in this book is listed in the Appendix, along with sources where it can be obtained. The majority of the components can be found in a starter kit from SparkFun, but most electronic starter kits will provide a lot of what you will need.

In many of the "how-tos," there will be a You Will Need section. This will refer to a code in the Appendix that explains where to get the component.

How to Use This Book

The book is organized into chapters, each of which has a theme. Within each chapter, most of the numbered sections contain a "how-to" on some topic of electronics.

The book contains the following chapters:

Chapter	Title	Description
Chapter 1	Getting Started	The book starts off by telling you where you can buy equipment and components, as well as things to hack. This chapter also deals with the basics of soldering and focuses on a project to hack an old computer fan to make a fume extractor for use while soldering.
Chapter 2	Theory and Practice	This chapter introduces electronic components—or at least the ones you are likely to use—and explains how to identify them and describes what they do. It also introduces a small amount of essential theory, which you will use over and over again.
Chapter 3	Basic Hacks	This chapter contains a set of fairly basic "hacking" how-tos, introducing concepts like using transistors with example projects. It includes hacking a "push light" to make it automatically turn on when it gets dark and how to control a motor using power MOSFETs.
Chapter 4	LEDs	In addition to discussing regular LEDs and how to use them and make them flash and so on, this chapter also looks at using constant current drivers for LEDs and how to power large numbers of LEDs and laser diode modules.
Chapter 5	Batteries and Power	This chapter discusses the various types of battery, both single use and rechargeable. It also covers how to charge batteries including LiPos. Automatic battery backup, voltage regulation, and solar charging are also explained.
Chapter 6	Hacking Arduino	The Arduino has become the microcontroller board of choice for electronics hackers. Its open-source hardware design makes using a complex device like a microcontroller very straightforward. The chapter gets you started with the Arduino and includes a few simple how-tos, like controlling a relay, playing sounds, and controlling servo motors from an Arduino. It also covers the use of Arduino expansion shields.

Chapter	Title	Description
Chapter 7	Hacking with Modules	When you want to make something, you can often use readymade modules at least for part of the project. Modules exist for all sorts of things, from wireless remotes to motor drivers.
Chapter 8	Hacking with Sensors	Sensor ICs and modules are available for sensing everything from gas to acceleration. In this chapter, we explore a good range of them and explain how to use them and connect some of them to an Arduino.
Chapter 9	Audio Hacks	This chapter has a number of useful how-tos relating to electronics and sound. It includes making and adapting audio leads, as well as audio amplifiers, and discusses the use of microphones.
Chapter 10	Mending and Breaking Electronics	Mending electronics and scavenging useful parts from dead electronics are a worthy activity for the electronics hacker. This chapter explains how to take things apart and sometimes put them back together again.
Chapter 11	Tools	The final chapter of the book is intended as a reference to explain more about how to get the most out of tools such as multimeters and lab power supplies.

HACKING ELECTRONICS

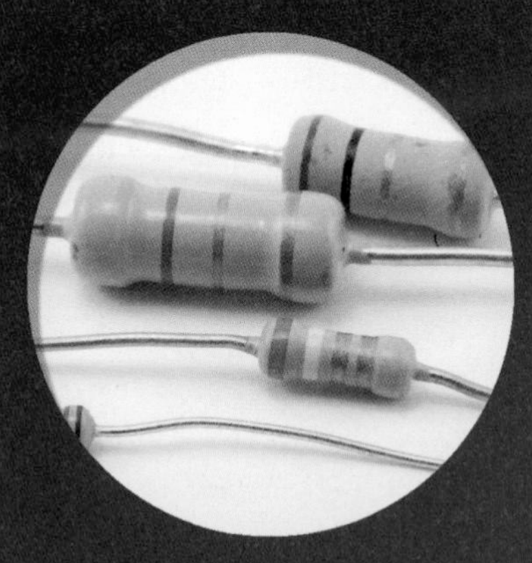

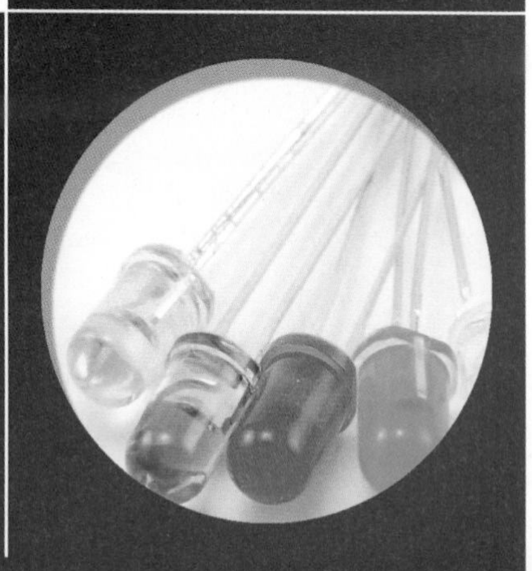

1

Getting Started

In this first chapter, we will investigate some of the tools and techniques needed to hack electronics. We will start with a little soldering, and wire up an old computer fan to help keep the solder fumes out of our lungs.

As it says in the title, this book is all about "hacking electronics." The word "hacking" has come to mean many things. But in this book, "hacking" means "just do it!" You don't need a degree in electronic engineering to create or modify something electronic. The best way to learn is by having a go at it. You will learn as much from your mistakes as from your successes.

As you start to make things and experiment, you will likely want to understand more of the theory behind it all. Traditional electronics textbooks are pretty terrifying unless you have a good grasp of complex mathematics. This book strives to, above all else, enable you to do things first and worry about the theory later.

To get started, you will need some tools, and also find out where to get components and parts to use in your projects.

Getting Stuff

In addition to buying components and tools, there are lots of low-cost and interesting electronic consumer items that can be hacked and used for new purposes, or that can act as donors of interesting components.

Buying Components

Most component purchases happen on the Internet, although there are local electronic stores like RadioShack (in the U.S.) and Maplin (in the UK) where you can buy components. At traditional brick-and-mortar stores like those, the product range is often limited and the prices can be on the high side. They do, after all, have a shop to pay for. These stores are invaluable, however, on the odd occasion where you need something in a hurry. Perhaps you need an LED because you accidentally destroyed one, or maybe you want to look at the enclosures they sell for projects.

Sometimes it's just nice to hold a box or look at tools for real, rather than trying to size them up from pictures on a web site.

As you get into electronics, you will likely gradually accumulate a set of components and tools that you can draw from when you start a new project. Components are relatively cheap, so when I need one of something, I generally order two or three or even five if they are cheap, enough that I have extras on hand that can be used another time. This way, you will often find that when you start to work on something, you actually have pretty much everything you need already.

Component buying really depends on where you are in the world. In the U.S., Mouser and DigiKey are the largest suppliers of electronic components to the hobby electronics market. In fact, both of these suppliers sell worldwide. Farnell also supplies pretty much anything you could want, anywhere in the world.

When it comes to buying ready-made electronics modules for your projects, the SparkFun, Seeed Studio, Adafruit, and ITead Studio web sites can help. All have a wide range of modules, and much enjoyment can be had simply from browsing their online catalogs.

Nearly all the components used in this book have part codes for one or more of the suppliers I just mentioned. The only exceptions are for a few unusual modules that are better to buy from eBay.

There is also no end to the electronic components available on online auction sites, many coming direct from countries in the far east and often at extremely low prices. This is frequently the place to go for unusual components and things like laser modules and high-power LEDs that can be expensive in regular component suppliers. They are also very good for buying components in bulk. Sometimes these components are not grade A, however, so read the descriptions carefully and don't be disappointed if some of the items in the batch are dead-on-arrival.

Where to Buy Things to Hack

The first thing to consider, now that you are into hacking electronics, is an effect that your household and friends will have on you. You will become the recipient of dead electronics. But keep an eye open in your new role as refuse collector. Sometimes these "dead" items may actually be candidates for straightforward resurrection.

Another major source of useful bits is the dollar/pound/euro (delete as appropriate) store. Find the aisle with the electronic stuff: flashlights, fans, solar toys, illuminated cooling laptop bases, and so on. It's amazing what can be bought for a single unit of currency. Often you will find motors and arrays of LEDs for a lower price than you would the raw components from a conventional supplier.

Supermarkets are another source of cheap electronics that can be hacked. Good examples of useful gadgets are cheap powered computer speakers, mice, power supplies, radio receivers, LED flashlights, and computer keyboards.

A Basic Toolkit

Don't think you are going to get through this chapter without doing some soldering. Given this, you will need some basic tools. These do not have to be expensive. In fact, when you are starting out on something new, it's a good idea to learn to use things that are inexpensive, so it doesn't matter if you spoil them. After all, you wouldn't learn the violin on a Stradivarius. Plus, what will you have to look forward to if you buy all your high-end tools now!

Many starter toolkits are available. For our purposes, you will need a basic soldering iron, solder, a soldering iron stand, some pliers, snips, and a screwdriver or two. SparkFun sells just such a kit (SKU TOL-09465), so buy that one or look for something similar.

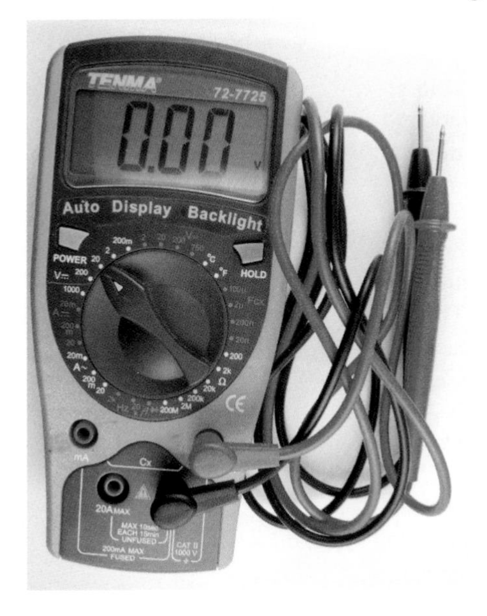

FIGURE 1-1 A digital multimeter

You will also need a multimeter (Figure 1-1). I would suggest a low-cost digital multimeter (don't even think of going above USD 20). Even if you end up buying a better one, you will still end up using the other one since it's often useful to measure more than one thing at a time. The key things you need are DC Volts, DC current, resistance, and a continuity test. Everything else is fluff that you will only need once in a blue moon. Again, look for something similar to this model from SparkFun (SKU TOL-09141) or the slightly higher specification meter shown in Figure 1-1.

Solderless breadboards (Figure 1-2) are very useful for quickly trying out designs before you commit them to solder. You poke the leads of components into the sockets, and metal clips behind the holes connect all the holes on a row together. They are not expensive (see T5 in the Appendix).

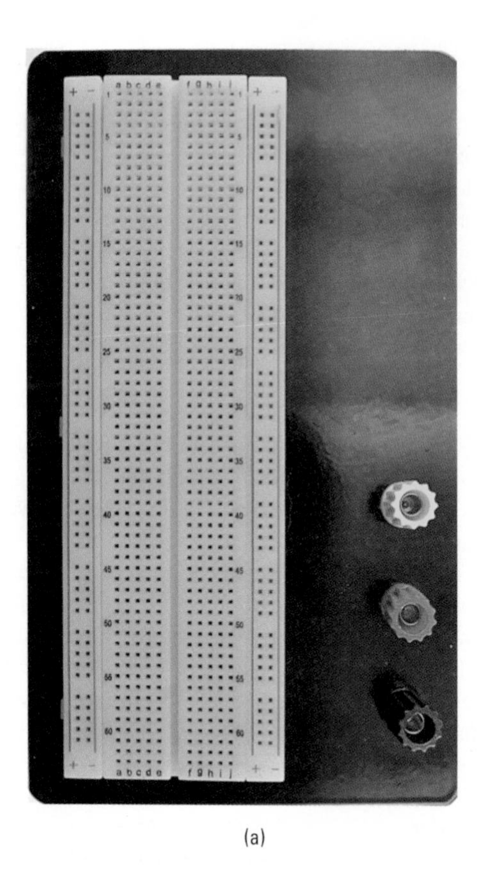

(a)

(b)

FIGURE 1-2 Solderless breadboard

You will also need some solid core wire in different colors (T6) to make bridging connections on the breadboard. Another good idea is to buy special-purpose jumper wires with little plugs on the end—although these are useful, they are by no means essential.

Breadboard comes in all shapes and sizes, but a big one is probably most useful. Where I use solderless breadboard in the book, I use the one specified in T5 in the Appendix. This has 63 rows by 2 columns with two supply strips down each side (Figure 1-2a). It is also mounted on an aluminum base with rubber feet to stop it moving about on the table. This is a very common size of breadboard and most suppliers will have something similar.

Figure 1-2b shows how the conductive strips are arranged underneath the plastic top surface of the board. All the holes that share a common gray area beneath are connected together in rows of five connectors. The long strips down each side are used for the power supply to the components. One positive and one negative. They are color-coded red and green.

How to Strip a Wire

Let's start with some basic techniques you need to know when hacking electronics. Perhaps the most basic of these is stripping wire.

You Will Need

Quantity	Item	Appendix Code
	Wire to be stripped	T9 or scrap
1	Pliers	T1
1	Snips	T1

Whenever you hack electronics, there is likely to be some wire involved, so you need to know how to use it. Figure 1-3 shows a selection of commonly used types of wire, set beside a matchstick to give them perspective.

On the left, next to the matchstick, are three lengths of solid-core wire, sometimes called hookup wire. This is mostly used with solderless breadboard, because being made of a single core of wire inside plastic insulation, it will eventually break if it is bent. Being made of a single strand of wire does mean it is much easier to push into sockets when prototyping since it doesn't bunch up like multi-core wire.

When using it with breadboard, you can either buy already-stripped lengths of wire in various colors as a kit (see Appendix, T6) or reels of wire that you can cut to the lengths you want yourself (see Appendix, T7, T8, T9). It is useful to have at least

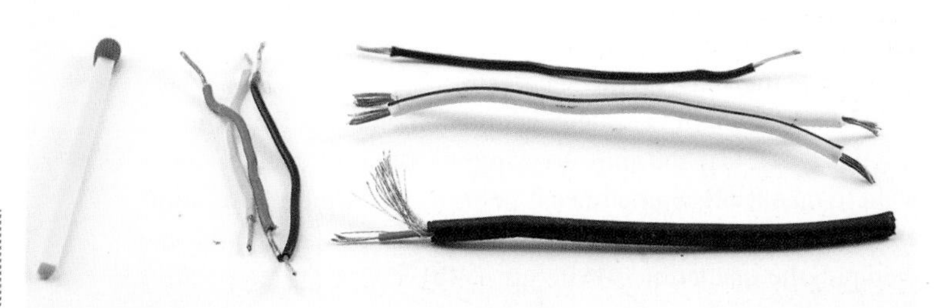

Figure 1-3 Common types of wire

three colors: red, yellow, and black are a good choice. It makes it easier to see how a project is connected up if you use red for the positive power supply, black for negative, and yellow for any other wires needed.

The top right of Figure 1-3 shows a length of multi-core wire, as well as some twin-strand multi-core wire. Multi-core wire is used when connecting up modules of a project. For instance, the wires to a loudspeaker from an amplifier module might use some twin, multi-core wire. It's useful to have some of this wire around. It is easily reclaimed from broken electronic devices, and relatively cheap to buy new (see Appendix, T10 and T11).

The wire at the bottom right of Figure 1-3 is screened wire. This is the type of wire you find in audio and headphone leads. It has an inner core of multi-core insulated wire surrounded by a screened wire on the outside. This type of wire is used where you don't want electrical noise from the environment such as mains hum (60 Hz electrical noise from 110V equipment) to influence the signal running through the central wire. The outer wire screens the inner wire from any stray signals and noise. There are variations of this where there is more than one core surrounded by the screening—for example, in a stereo audio lead.

Insulated wire is of no use to us unless we have a way of taking some of the insulation off it at the end, as this is where we will connect it to something. This is called "stripping" the wire. You can buy special-purpose wire strippers for this, which you can adjust to the diameter of the wire you want to strip. This implies that you know the width of the wire, however. If you are using some wire that you scavenged from a dead electronic appliance, you won't know the width. Having said that, with a bit of practice you will find you can strip wire just as well using a pair of pliers and some wire snips.

Both of these are essential tools for the electronics hacker. Neither tool needs to be expensive. In fact, snips tend to get notches in them that make them annoying to use, so a cheap pair (I usually pay about USD 2) that can be replaced regularly is a good idea.

Figures 1-4a and 1-4b show how to strip a wire with pliers and snips. The pliers are used to hold things still with a firm grip, while the snips do the actual stripping.

Grip the wire in the pliers, about an inch away from the end (Figure 1-4a). Use the snips to grip the insulation where you want to take it off. Sometimes it helps to just nip the insulation all the way around before gripping it tightly with the snips, and then pull the insulation off (Figure 1-4b).

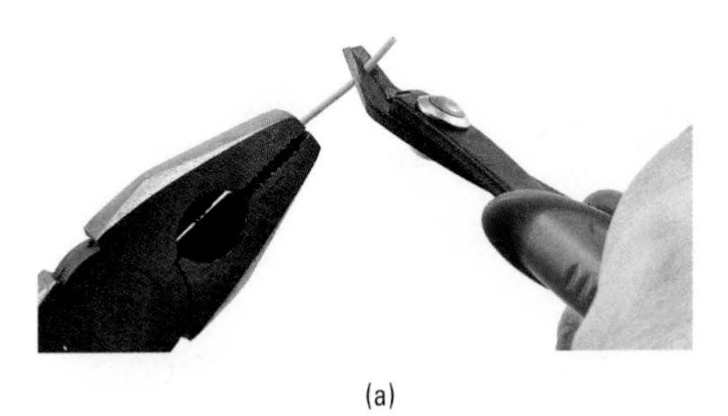

(a)

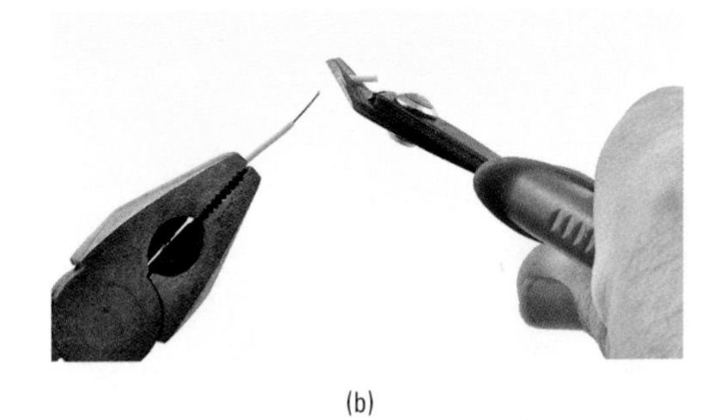

(b)

FIGURE 1-4 Stripping wire

For longer lengths of wire, you can just wrap the wire around your finger a few times instead of using pliers.

This takes a bit of practice. Sometimes you will have the snips grip it too tightly and accidentally cut the wire all the way through, while other times you won't grip it hard enough with the snips and the insulation will stay in place or stretch. Before attempting anything important, practice with an old length of wire.

How to Join Wires Together by Twisting

It is possible to join wires without soldering. Soldering is more permanent, but sometimes this technique is good enough.

One of the simplest ways of joining wires is to simply twist the bare ends together. This works much better for multi-core wire than the single-core variety, but if done properly with the single-core, it will still make a reliable connection.

You Will Need

To try out joining two wires by twisting (there is slightly more to it than you might expect), you will need the following.

Quantity	Item	Appendix Code
2	Wires to be joined	T10
1	Roll of PVC insulating tape	T3

If you need to strip the wires first to get at the copper, refer back to the section "How to Strip a Wire."

Figures 1-5a thru 1-5d show the sequence of events in joining two wires by twisting them.

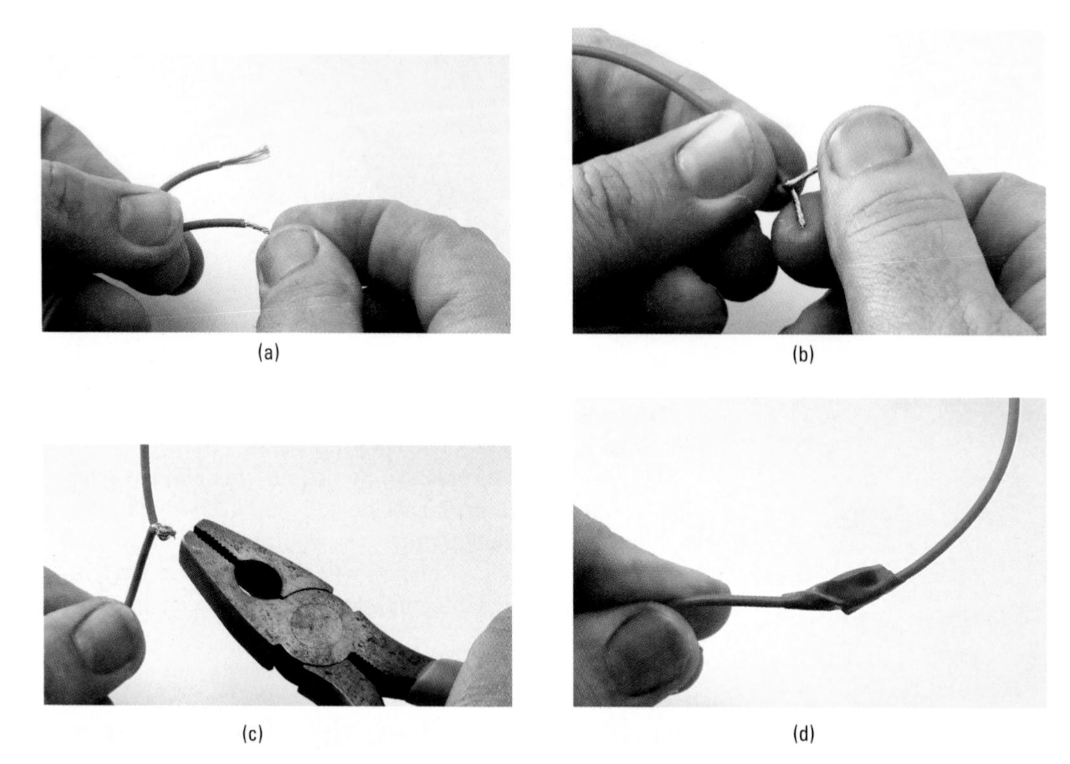
(a) (b) (c) (d)

FIGURE 1-5 Joining wires by twisting

First, twist the strands of each wire up clockwise (Figure 1-5a). This just tidies up any straggling strands of the multi-core wire. Then, twist together the two pre-twisted wires (Figure 1-5b) so they are both twisting around each other. Try to avoid the situation where one of the wires twists around the second, while the second remains straight. If it does this, it is very easy for the first wire to just slip off the second. Next, twist the joined wires up into a neat little knot (Figure 1-5c). Note that a pair of pliers may be easier to use when making the knot, especially if the wire is on the thick side. Lastly, cover the joint with four or five turns of PVC insulating tape (Figure 1-5d).

How to Join Wires by Soldering

Soldering is the main skill necessary for hacking electronics.

Safety

I don't want to put you off, but … be aware that soldering involves melting metal at very high temperatures. Not only that, but melting metal that's coupled with noxious fumes. It is a law of nature that anyone who has a motorbike eventually falls off it, and anyone who solders will burn their fingers. So be careful and follow these safety tips:

- Always put the iron back in its stand when you are not actually soldering something. If you leave it resting on the bench, sooner or later it will roll off. Or you could catch the wires with your elbow and if it falls to the floor, your natural reflex will be to try and catch it—and chances are you will catch the hot end. If you try and juggle it in one hand, while looking for something or arranging some components ready to solder, sooner or later you will either solder your fingers or burn something precious.
- Wear safety glasses. Blobs of molten solder will sometimes flick up, especially when soldering a wire or component that is under tension. You do not want a blob of molten solder in your eye. If you are long-sighted, magnifying goggles may not look cool, but they will serve the dual purpose of protecting your eyes and letting you see properly.
- If you do burn yourself, run cold water over the burned skin for at least a minute. If the burn is bad, seek medical attention.
- Solder in a ventilated room, and ideally set up a little fan to draw the fumes away from you and the soldering iron. Preferably have it blowing out of a window. A fun little project to practice your wire joining skills on is making a fan using an old computer (see the section "How to Hack a Computer Fan to Keep Soldering Fumes Away").

You Will Need

To practice joining some wires with solder, you will need the following items.

Quantity	Item	Appendix Code
2	Wires to be joined	T10
1	Roll of PVC insulating tape	T3
1	Soldering kit	T1
1	Magic hands (optional)	T4
1	Coffee mug (essential)	

Magic hands are a great help during soldering because they solve the problem that, when soldering, you really need three hands: one to hold the iron, one to hold the solder, and one to hold the thing or things you are trying to solder. You generally use the magic hands to hold the thing or things you are trying to solder. Magic hands are comprised of a small weighted bracket with crocodile clips that can be used to hold things in place and off the work surface.

An alternative that works well for wires is to bend them a little so that the end you are soldering will stick up from the workbench. It usually helps to place something heavy like a coffee mug on the wire to keep it from moving.

Soldering

Before we get onto the business of joining these two wires, let's have a look at soldering. If you haven't soldered before, Figures 1-6a thru 1-6c show you how it's done.

1. Make sure your soldering iron has fully heated up.
2. Clean the tip by wiping it on the damp (not sopping wet) sponge on the soldering iron stand.

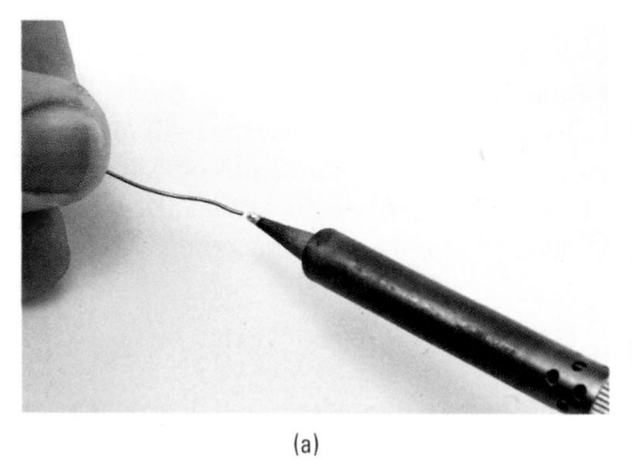

(a)

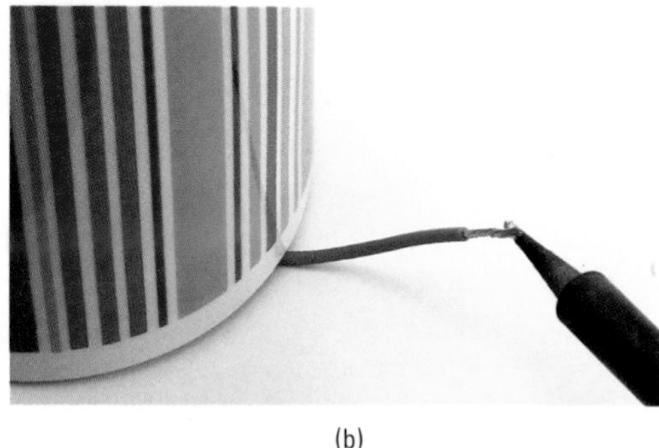

(b)

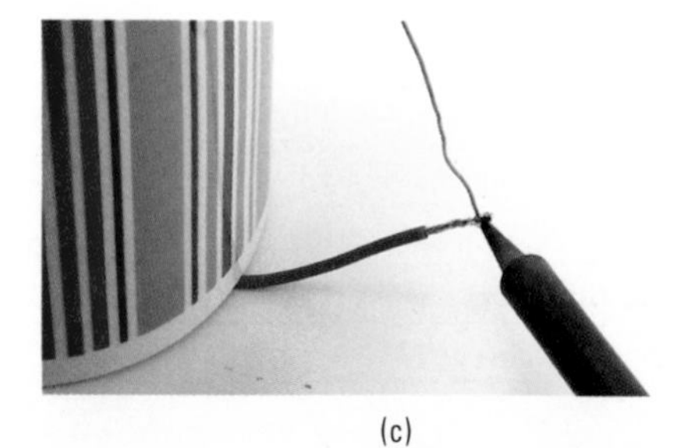

(c)

Figure 1-6 Soldering—tinning a wire

3. Touch a bit of solder onto the tip of the iron to "tin" it (see Figure 1-6a). After you have done this, the tip should be bright and shiny. If the solder doesn't melt, then your iron probably isn't hot enough yet. If the solder forms into a ball and doesn't coat the tip of the iron, the tip of it may be dirty, so wipe it on the sponge and try again.
4. Hold the soldering iron to the wire and leave it there for a second or two (Figure 1-6b).
5. Touch the solder to the wire near the soldering iron. It should flow into the wire (Figure 1-6c).

Soldering is something of an art. Some people are naturally very neat at soldering. So do not worry if your results are a bit blobby at first. You will get better. The main thing to remember is that you heat up the item you want to solder and only apply the solder when that thing is hot enough for the solder to melt onto it. If you are struggling, it sometimes helps to apply the solder to the spot where the soldering iron meets the thing being soldered.

The following section offers a bit more soldering practice for you—in this case, by soldering wires together.

Joining Wires

To join two wires with solder, you can use the same approach described in the section "How to Join Wires Together by Twisting" and then flow solder into the little knot. An alternative way—that makes for a less lumpy joined wire—is illustrated in Figures 1-7a thru 1-7d.

1. The first step is to twist each end. If it is multi-core wire (a), tin it with solder as shown in Figure 1-7a.
2. Hold the wires side by side and heat them with the iron (see Figure 1-7b). Note the chopstick technique of holding both the second wire and the solder in one hand.
3. Introduce the solder to the wires so they join together into one wire and look something like that shown Figure 1-7c.
4. Wrap the joint in three or four turns of insulating tape—half an inch is probably enough (see Figure 1-7d).

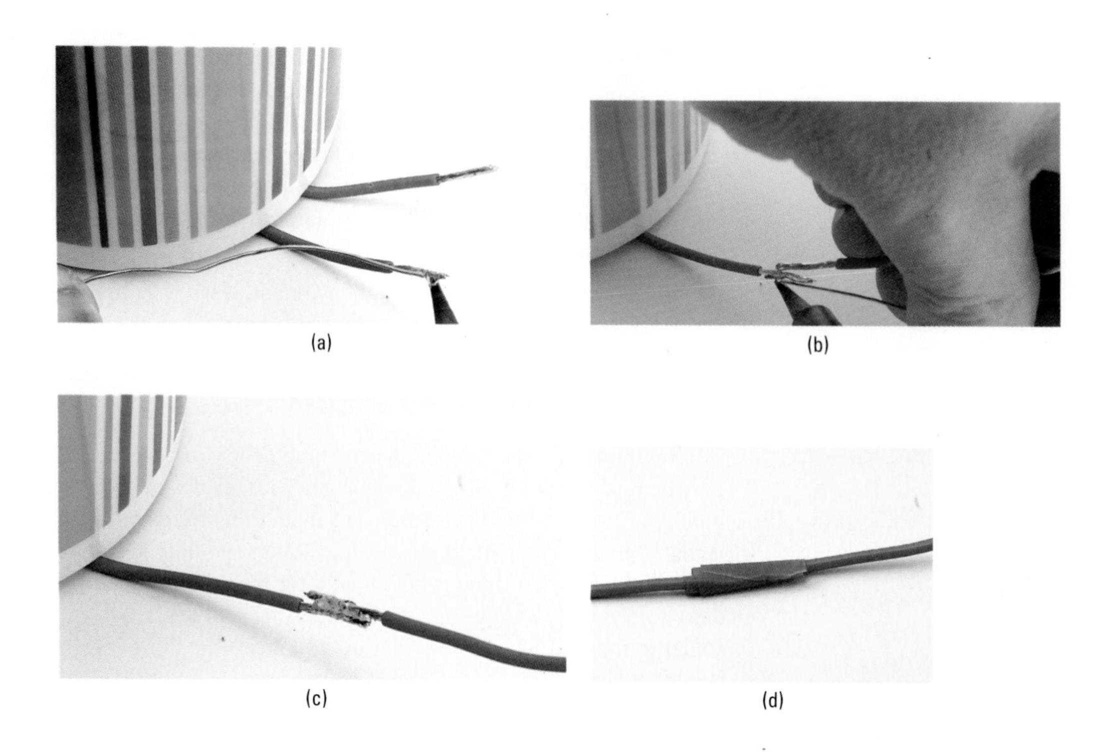
(a) (b) (c) (d)

FIGURE 1-7 Joining wires by soldering

How to Test a Connection

For the joints that we have made in the section "How to Join Wires by Soldering," it is fairly obvious that they are connected. However, especially with solid-core wire, it is not uncommon for the wire core to break somewhere under the insulation. If you own an electric guitar, you will probably be familiar with the problem of a broken guitar lead.

You Will Need

Quantity	Item	Appendix Code
1	Multimeter	T2
1	Connections to be tested	

Nearly all multimeters have a "Continuity" mode. When set in this useful mode, the multimeter will beep when the leads are connected to each other.

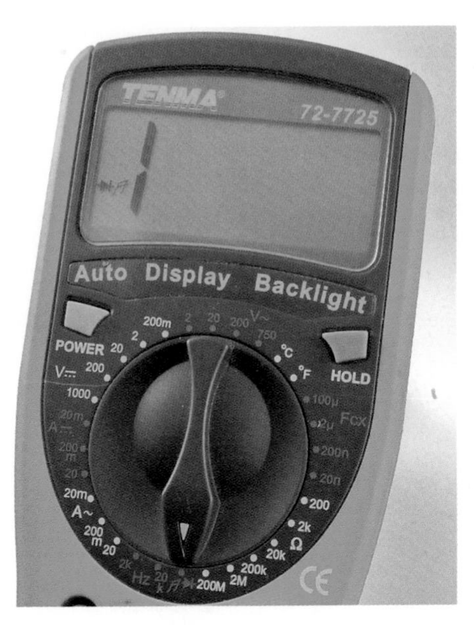

Figure 1-8 A multimeter in Continuity mode

Set your multimeter to "Continuity mode," and then try touching the leads together. Now take a length of wire and try touching the multimeter leads to each end of the wire (Figure 1-8). The buzzer should sound if the wire is okay.

You can use this technique on circuit boards. If you have an old bit of circuit board from something, try testing between the soldered connections on the same track (Figure 1-9).

If there is no connection where you would expect there to be a connection, then there may be a "dry joint," where the solder hasn't flowed properly or there is a crack in the track on the circuit board (this sometimes happens if the board gets flexed).

A dry joint is easily fixed by just applying a bit of solder and making sure it flows properly. Cracks on a circuit board can be fixed by scraping away some of the protective lacquer over the track and then soldering up the split in the track.

Figure 1-9 Testing a circuit board

How to Hack a Computer Fan to Keep Soldering Fumes Away

Solder fumes are unpleasant and bad for you. If you can sit by an open window while you solder, then great. If not, then this is a good little construction project to enhance your electronics hacking skills (Figure 1-10).

Okay, so it's not going to win any awards for style, but attached to my work light (which is always close to whatever I am soldering), the fumes will at least be directed away from my face.

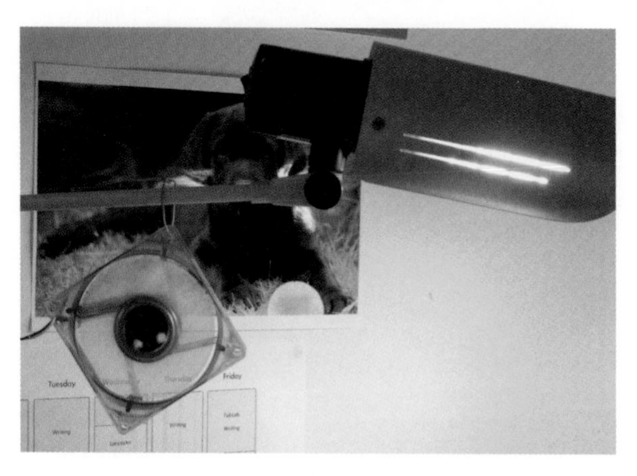

FIGURE 1-10 A homemade fume extractor

You Will Need

Quantity	Item	Appendix Code
1	Soldering equipment	T1
1	An old computer fan (two-lead)	
1	12V power supply	M1
1	SPST switch	K1

Construction

Figure 1-11 shows the schematic diagram for this mini-project.

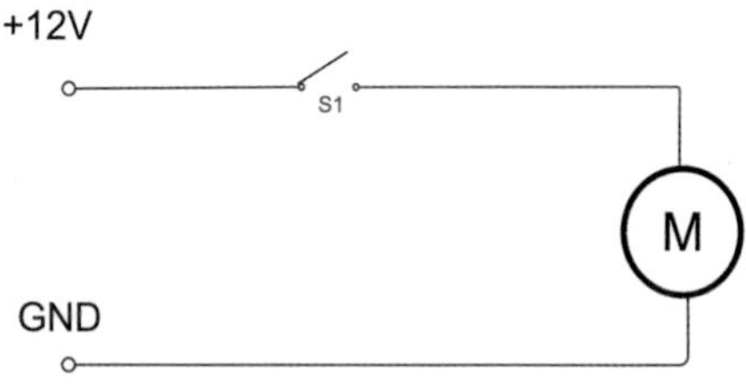

FIGURE 1-11 The schematic diagram for the fume extractor

Newcomers to electronics often view schematic diagrams like this with suspicion, thinking it better just to show the components as they actually are, with wires where wires need to be—just like in Figure 1-12. It is worth learning how to read a schematic diagram. It really isn't that hard and in the long term it will pay dividends. Not least because of the vast number of useful circuit diagrams published on the Internet. It's a bit like being able to read music. You can get so far playing by ear, but there are more options if you can read and write musical notation.

So, let's examine our schematic diagram. Over on the left we have two labels that say "+12V" and "GND." The first is the

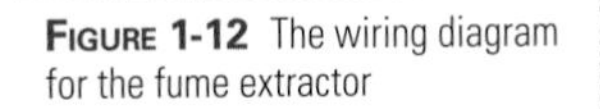
Figure 1-12 The wiring diagram for the fume extractor

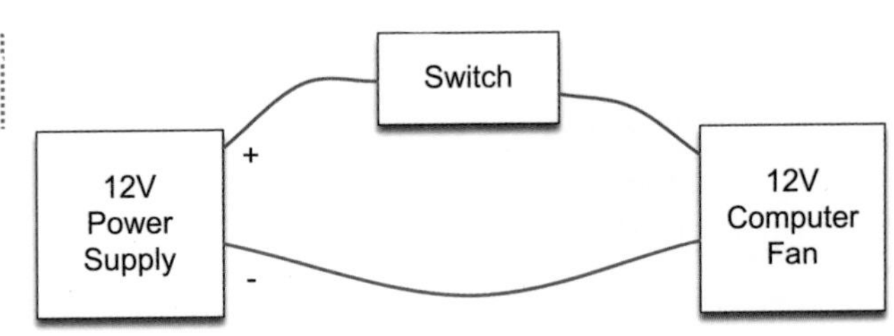

12V positive supply from the 12V power supply. GND actually refers to the negative connection of the power supply. GND is short for "ground" and just means zero volts. Voltage is relative, so the 12V connection of the power supply is 12V above the other connection (the GND connection). We will learn more about voltage in the next chapter.

Moving toward the right, we have a switch. This is labeled "S1," and if we had more than one switch in a schematic, they would be labeled "S2," "S3," and so on. The symbol for a switch shows how it operates. When the switch is turned to the on position, its two connections are connected together, and when it is in the off position, they aren't. It's as simple as that.

The switch is just controlling the supply of electricity to the motor of the fan (M) as if it were a faucet.

Step 1. Strip the Power Supply Leads

We have a power supply and we are going to cut the plug off the end of it and strip the wires (see the section "How to Strip a Wire"). Before you cut off the plug, make sure the power supply is NOT plugged in. Otherwise, if you snip both wires at the same time, the cutters will probably short the two connections together, which may damage the power supply.

Step 2. Identify the Power Supply Lead Polarity

Having cut the wires, we need to know which one is the positive one. To do this, let's use a multimeter. Set the multimeter to its 20V DC range. Your multimeter will probably have two voltage ranges, one for AC and one for DC. You need to use the DC range. This is often marked by a solid line above a dotted line. The AC range will either be marked as AC or have a picture of a little sine wave next to it. If you select AC instead of DC, it will not damage the meter, but you will not get a meaningful reading. (See Chapter 11 if you need more information on multimeters.)

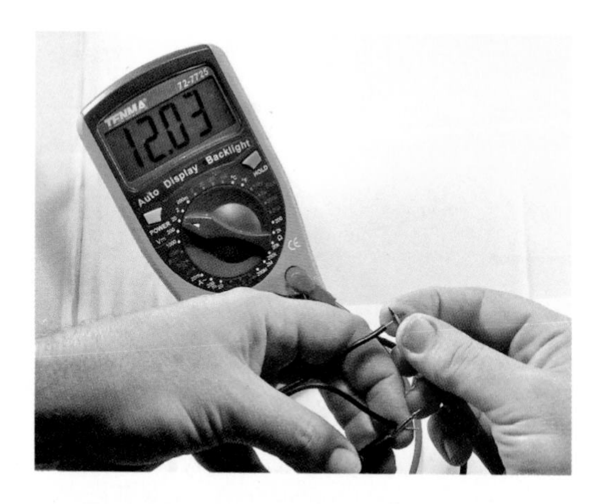

FIGURE 1-13 Using a multimeter to find the power supply polarity

First making sure that the stripped leads from the power supply are not touching, plug the power supply in and turn it on.

Now touch the two test leads from the multimeter to the leads from the power supply (Figure 1-13). If the number on the multimeter is not negative, then the red test lead of the multimeter is connected to the positive lead. Mark the lead in some way (I tied a knot in it). If the multimeter shows a negative voltage, then the leads are swapped over, so tie a knot in the power supply lead connected to the black test lead of the multimeter—in this case, this is the positive lead from the power supply.

Step 3. Connect the Negative Leads Together

Unplug your power supply. You should never solder anything that is powered up.

Cut any plug off the end of your computer fan and strip the two wires. Mine had one black (negative) and one yellow (positive) lead. Three lead fans are more complex and should be avoided. If you get the leads the wrong way around, no harm will befall you. The fan will just rotate in the opposite direction.

We are now going to join the negative lead of the fan to the negative lead (no knot) of the power supply (Figure 1-14).

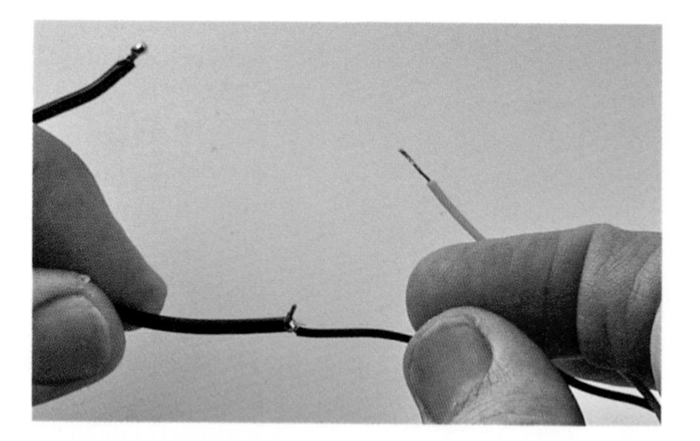

FIGURE 1-14 Connecting the negative leads together

FIGURE 1-15 Connecting the positive lead to the switch

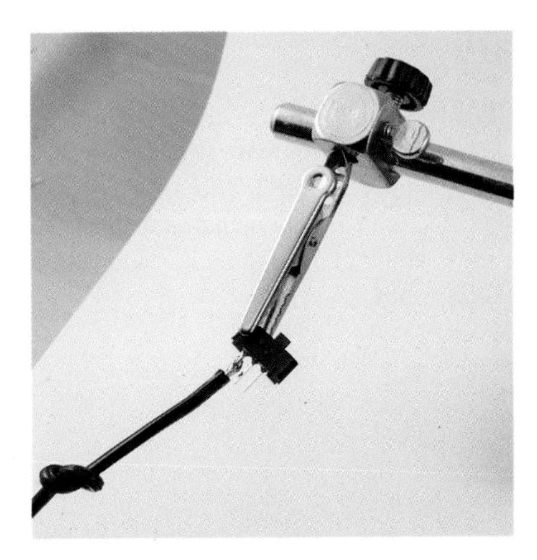

Step 4. Connect the Positive Lead to the Switch

Solder the positive lead from the power supply to one of the outer connections on the switch (it doesn't matter which). (See Figure 1-15.) It will help to tin the switch connection with a little solder before you start.

Finally, connect the remaining lead from the fan to the center connection of the switch (see Figure 1-16).

Step 5. Try It Out

Wrap the bare connections with insulating tape, plug it in, turn it on, and presto! When you flick the switch, the fan should come on.

FIGURE 1-16 Connecting the fan to the switch

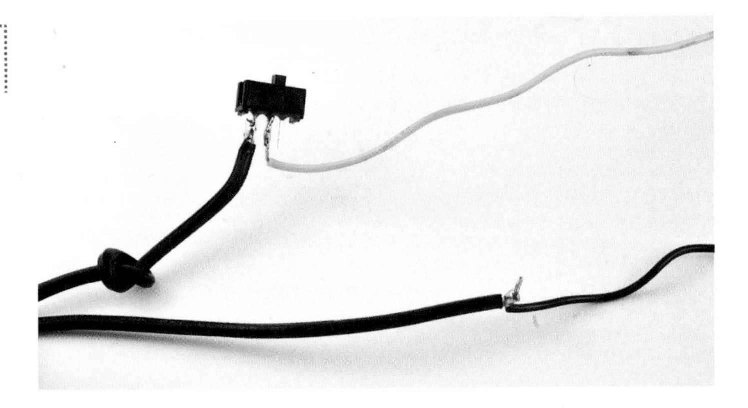

Summary

Now that we have the basics and are confident about a bit of soldering and dealing with wires and switches, we can now move on to Chapter 2. There, we will start looking at a few electronic components, as well as some of the basic ideas you will need to understand to successfully hack electronics.

2

Theory and Practice

There are a few fundamentals that will help us get the most out of our electronics. I have no intention of overloading you with theory, so you may find you come back to this chapter as and when you need to. But before we start on any theory, let's look at getting together some of the components we will use.

How to Assemble a Starter Kit of Components

In Chapter 1, we assembled a few tools and did some soldering. The only thing we made used a scavenged computer fan, an off-the-shelf power supply, and a switch.

Certain components you will find that you use over and over again. To get yourself a basic stock of components, I recommend you buy a starter kit. SparkFun sells such a kit (see the Appendix, K1), but it does not contain any resistors, so you will need to buy a resistor set, too (K2). Once you have these, you will have a useful collection of components that should cover 80 percent of what you need.

Other suppliers sell starter kits, and although none of them will contain everything you need for this book, most will give you a very good starting point.

You Will Need

The SparkFun Starter Kit contains the following items, and the items used directly in this book are marked with a *, so if buying an alternative kit, look for one that has the majority of these components. Also see the Appendix for a list of other components used in the book.

Quantity	Item	Quantity	Item
10	0.1uF capacitor *	3	20-pin male header *
5	100uF capacitor *	3	Mini power switch *
5	10uF capacitor *	2	Push buttons *
5	1uF capacitor	1	10k trimpot *
5	10nF capacitor	2	LM358 OpAmp
5	1nF capacitor	2	3.3V regulator
5	100pF capacitor	2	5V regulator *
5	10pF capacitor	1	555 timer *
5	1N4148 diode	1	Green LED *
5	1N4001 diode *	1	Yellow LED *
5	2N3906 PNP transistor	1	Red LED *
5	2N3904 NPN transistor *	1	7-segment red LED
3	20-pin female header	1	Mini photocell *

The separate SparkFun resistor kit (K2) contains resistors of the following values:

0Ω, 1.5Ω, 4.7Ω, 10Ω, 47Ω
110Ω, 220Ω, 330Ω, 470Ω, 680Ω
1kΩ, 2.2kΩ, 3.3kΩ, 4.7kΩ, 10kΩ
22kΩ, 47kΩ, 100kΩ, 330kΩ, 1MΩ

How to Identify Electronic Components

So, what have we just bought here? Let's go through the components in the SparkFun starter kits and explain what they do, starting with the resistors.

Resistors

Figure 2-1 shows an assortment of resistors. Resistors come in different sizes to be able to cope with different amounts of power. High-power resistors are physically big to cope with the heat they produce. Since "parts getting hot" is generally a bad thing in electronics, we will mostly avoid that. Nearly all of the time we can use the 0.25-watt resistors as provided in the SparkFun kit, which are perfect for general use.

As well as having a maximum power rating, resistors also have a "resistance." As the word suggests, resistance is actually

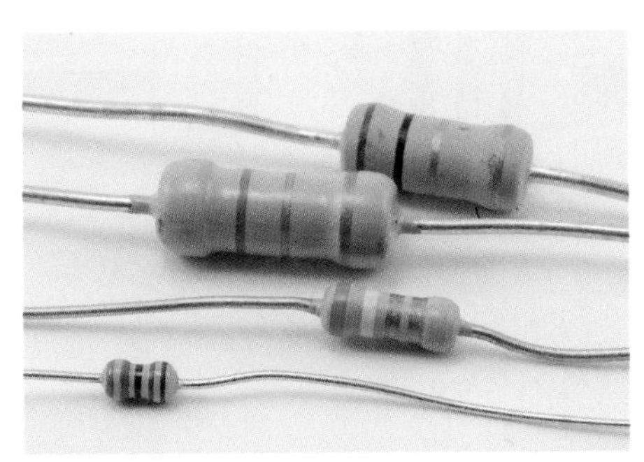

FIGURE 2-1 Assorted resistors

resistance to the flow of current. So a high-resistance resistor will not allow much current to flow, while a low-value resistor will allow lots of current to flow.

Resistors are the most commonly used component you can find. Since we will be using them a lot, we will go into greater detail on the subject in the section "What Are Current, Resistance, and Voltage?" later in this chapter.

Resistors have little stripes on them that tell you their value. You can learn to read the stripes (more in a moment on that) or you can avoid all of this by storing them in a bag or in the drawer of a component box with the value written on the box or bag. If in doubt, check the value with the resistance measurement feature of your multimeter.

However, an essential piece of geekiness is to know your resistor color-codes. Each color has a value, as shown next:

Color	Value
Black	0
Brown	1
Red	2
Orange	3
Yellow	4
Green	5
Blue	6
Violet	7
Gray	8
White	9
Gold	1/10
Silver	1/100

Gold and silver, as well as representing the fractions 1/10 and 1/100, are also used to indicate how accurate the resistor is. So gold is ±5% and silver is ±10%.

There will generally be three of these bands grouped together at one end of the resistor. This is followed by a gap, and then a single band at the other end of the resistor. The single band indicates the accuracy of the resistor value. Since none of the projects in this book require very accurate resistors, there is no need to select your resistors on the basis of accuracy.

Figure 2-2 shows the arrangement of the colored bands. The resistor value uses just the three bands. The first band is the first digit, the second the second digit, and the third "multiplier" band is how many zeros to put after the first two digits.

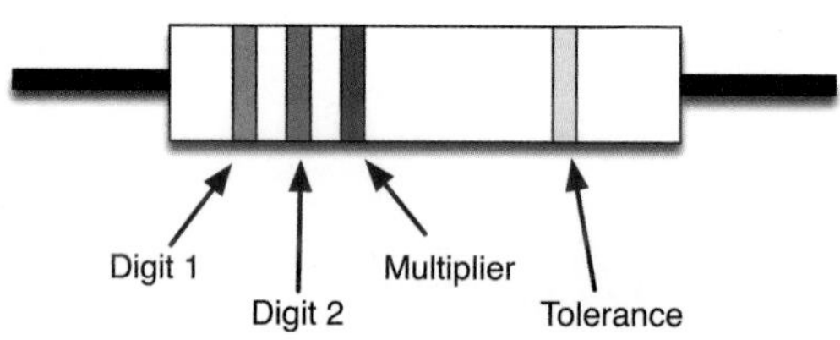

FIGURE 2-2 Resistor stripes

So, a 270Ω (*ohm*) resistor will have first digit 2 (red), second digit 7 (violet), and a multiplier of 1 (brown). Similarly, a 10KΩ resistor will have bands of brown, black, and orange (1, 0, 000).

In addition to fixed resistors, there are also variable resistors (a.k.a., potentiometers or pots). This comes in handy with volume controls, where turning a knob changes the resistance and alters the level of sound.

Capacitors

When hacking electronics, you will occasionally need to use a capacitor. Luckily, you do not need to know much about what they do. They are often used to head-off problems like the instability of a circuit or unwanted noise. Their use is often given a name like "decoupling capacitor" or "smoothing capacitor." There are simple rules you can follow about where you need a capacitor. These will be highlighted as we encounter them in later sections.

For the curious, capacitors store charge, a bit like a battery, but not much charge, and they can store the charge and release it very quickly.

Figure 2-3 shows a selection of capacitors.

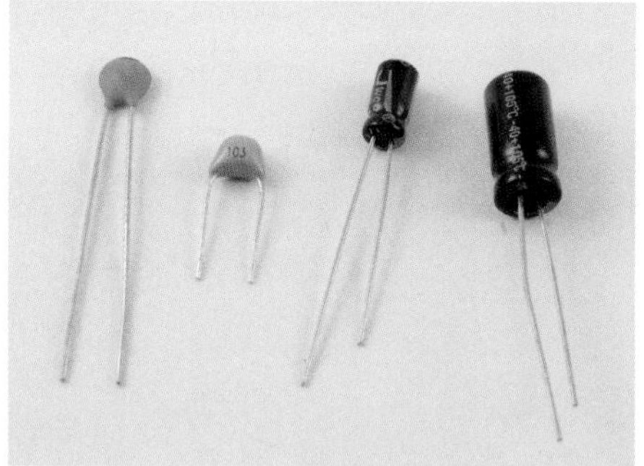

FIGURE 2-3 Assorted capacitors

If you look closely at the second capacitor from the left, you will see the number 103. This is actually the value of the capacitor in picofarads. The unit of capacitance is farad, but a 1F capacitor would be considered a huge capacitor, storing a great deal of charge. So, while such beasts do exist, everyday capacitors are either measured in nanofarads (nF = 1/1,000,000,000F) or microfarads (μF = 1/1,000,000F). You will also find capacitors in the picofarad range (pF = 1/1,000,000,000,000F).

Returning to 103. … Rather like resistors, this means 10 and then 3 zeros, in units of pF. So in this case that's 10,000pF or 10nF.

Larger capacitors, like those on the right of Figure 2-3, are called electrolytic capacitors. They are usually in the μF range and have their value written on their side. They also have a + and a – side, and unlike most other capacitors must be connected the right way around.

Figure 2-4 shows a large electrolytic, with value (1000µF) and its negative lead clearly indicated at the bottom of the figure. If the capacitor has one lead longer than the other, the longer one will normally be the positive lead.

The capacitor in Figure 2-4 also has a voltage written on it (200V). This is the capacitor's maximum voltage. So if you put more than 200V across its leads, it will fail. Big electrolytic capacitors like this have a reputation for failing spectacularly and may burst, spewing forth goo.

FIGURE 2-4 An electrolytic capacitor

Diodes

You will occasionally need to use diodes. They are kind of a one-way valve, only allowing current to flow in one direction. They are therefore often used to protect sensitive components from accidental reverse voltage that could damage them.

Diodes (Figure 2-5) have a stripe at one end. That end is called the cathode, while the other end is called the anode. We will hear more about diodes later.

As with resistors, the bigger the diode physically, the more power it can cope with before it gets too hot and expires. Ninety percent of the time, you will just be using one of the two diodes on the left-hand side of the figure.

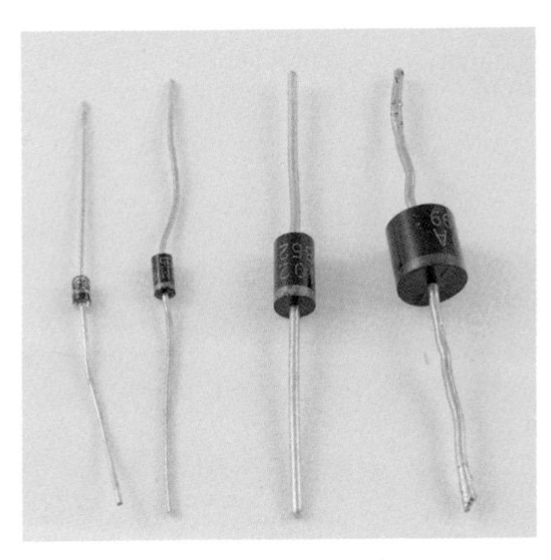

FIGURE 2-5 A selection of diodes

LEDs

LEDs light up, and generally look pretty. Figure 2-6 shows a selection of LEDs.

LEDs are a little sensitive, so you should not connect them directly to a battery. Instead you have to use a resistor to reduce the current flowing into the LED. If you do not do this, the LED will probably die almost instantly.

Later on, we will see how to select the right resistor for the job.

Just like regular diodes, LEDs have a positive and a negative lead (anode and cathode). The anode is the longer of the two leads. There is also usually a flat side to the LED case on the cathode side.

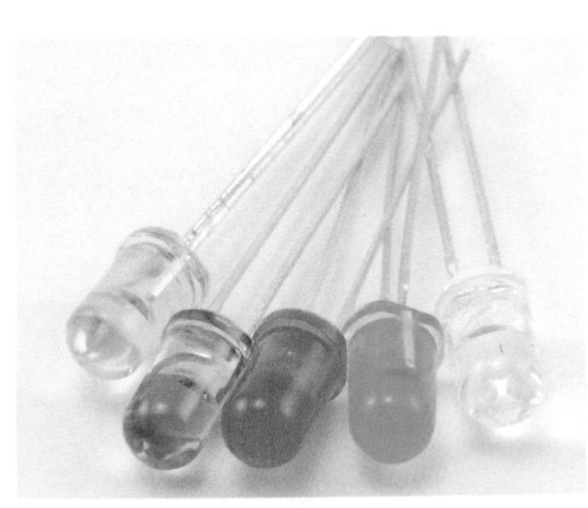

FIGURE 2-6 Assorted LEDs

As well as single LEDs, you also get LEDs in more complicated arrangements within a single package. Figure 2-7 shows some interesting-looking LEDs.

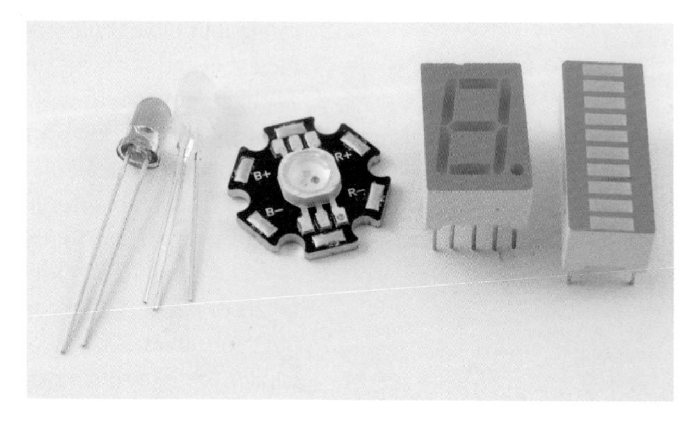

FIGURE 2-7 More LEDs

From left to right, these LEDs are an ultraviolet LED, an LED with both red and green LEDs in the same package, a high-power RGB (red, green, blue) LED that can be controlled to produce any color of light, a seven-segment LED display, and an LED bar graph display.

This is just a small selection of LED types. There are many others to choose from. In later sections, we will explore some of these more exotic LEDs.

Transistors

While transistors can be used in audio amplifiers and in many circumstances, for the casual electronics hacker, the transistor can be thought of as a switch. But rather than a switch controlled by a lever, it is a switch that switches a big current, yet is controlled by a small current.

Generally speaking, the physical size of the transistor (Figure 2-8) determines how big the current that it switches can be before it starts producing smoke.

FIGURE 2-8 Transistors

Of the transistors in Figure 2-8, the right-hand two are quite specialized and employed for high power use.

Generally, the rule for a component is that if it's ugly and has three legs, it's probably some kind of transistor.

Integrated Circuits

An integrated circuit (IC), or just "chip," is a load of transistors and other components printed onto silicon. The purpose of the IC varies wildly. It can be a microcontroller (mini-computer), or an entire audio amplifier, or a computer memory, or any one of thousands of other possibilities.

ICs make life easy, because as they say, often "there's a chip for that." Indeed, if there is something you want to make, there

FIGURE 2-9 Integrated circuits

may well be a chip for it already, and if there isn't, then there will probably be a general-purpose chip that takes you halfway there.

ICs look like bugs (Figure 2-9).

Other Stuff

There are so many other components out there, some of which are very familiar, such as batteries and switches. Others are less familiar and include potentiometers (variable resistors found in volume controls), phototransistors, rotary encoders, light dependent resistors, and so on. We will explore these as they arise later in the book.

Surface Mount Components

Let's touch a little on the subject of surface mount devices (SMDs). These components are just resistors, transistors, capacitors, ICs, and so on, but in tiny packages designed to be soldered onto the top surface of circuit boards by machines.

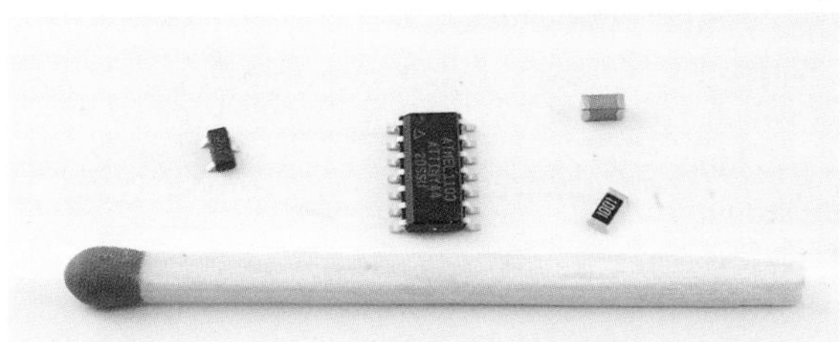

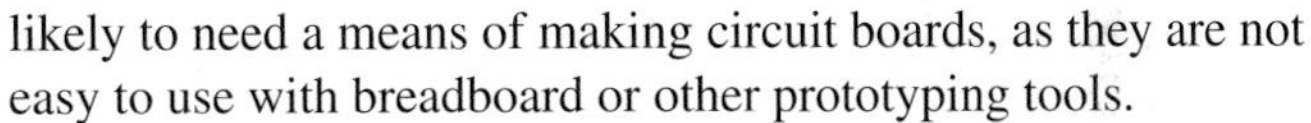

FIGURE 2-10 Surface mount components

Figure 2-10 shows a selection of SMDs.

The matchstick shows you just how small these devices are. It is perfectly possible to do surface mount soldering by hand, but you need a steady hand and a high-quality soldering iron. Not to mention a lot of patience. You are also likely to need a means of making circuit boards, as they are not easy to use with breadboard or other prototyping tools.

In this book, we mostly look at using the conventional "through-hole" components rather than SMDs. However, as your experience grows and you feel you might like working with SMDs, do not be afraid to try.

What Are Current, Resistance, and Voltage?

Voltage, current, and resistance are three properties that are fundamental to almost everything you will do in electronics. They are intimately related, and if you can master the relationship between them, you will be a wise hacker indeed.

Please take the time to read and understand this little bit of theory. Once you understand it, many other things should automatically fall into place.

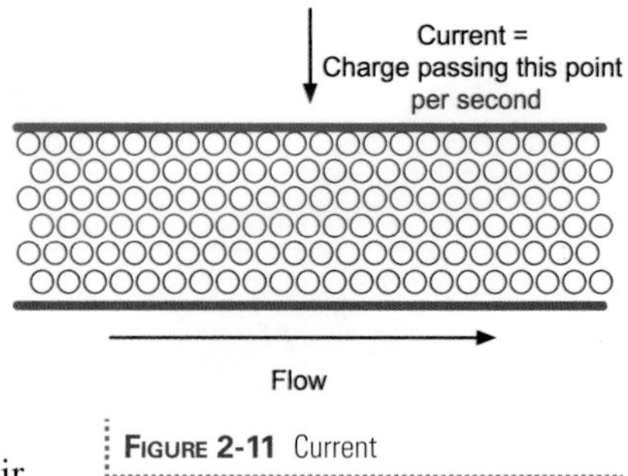

FIGURE 2-11 Current

Current

The problem with electrons is that you cannot see them, so you just have to imagine how they do things. I like to think of electrons as little balls flowing through pipes. Any physicists reading this will probably be clutching their heads or hurling this book to the floor in disgust now. But it works for me.

Each electron has a charge and it's always the same—lots of electrons, lots of charge, few electrons, and a little bit of charge.

Current, rather like the current in a river, is measured by counting how much charge passes you per second (Figure 2-11).

Resistance

A resistor's job is to provide resistance to the flow of current. So, if we keep thinking about our river, it is like a constriction in a river (Figure 2-12).

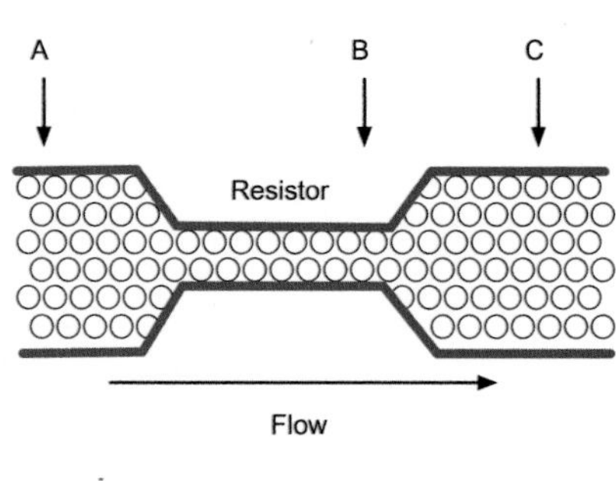

FIGURE 2-12 A resistor

The resistor has reduced the amount of charge that can pass by a point. And it doesn't matter which point you measure at (A, B, or C) because, if you look upstream of the resistor, the charge is hanging around waiting to move through the resistor. Therefore, less is moving past A per second. In the resistor (B), it's restricted.

The "speed" analogy does not really hold true for electrons, but one important point is that the current will be the same wherever you measure it.

Imagine what happens when a resistor stops too much current from flowing through an LED.

Voltage

Voltage is the final part of the equation (that we will come to in a minute). If we persist with the water-in-a-river analogy, then voltage is like the height that the river drops over a given distance (Figure 2-13).

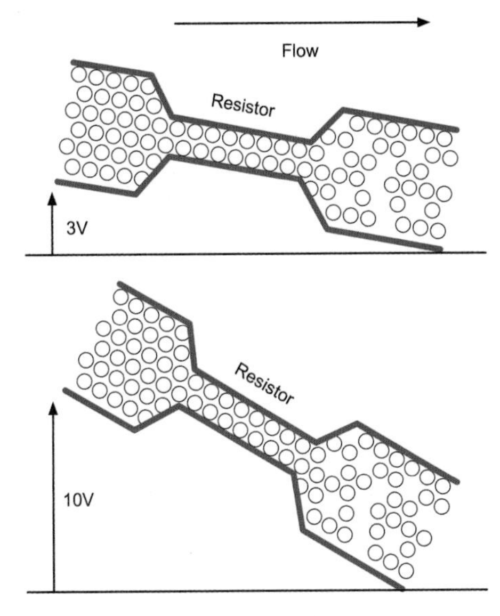

FIGURE 2-13 Voltage

As everyone knows, a river that loses height quickly flows fast and furious, whereas a relatively gently sloped river will have a correspondingly gentle current.

This analogy helps with the concept of voltage being relative. That is, it does not matter if the river is falling from 10,000 ft to 5,000 ft or from 5,000 ft to 0 ft. The drop is the same and so will be the rate of flow.

Ohm's Law

Before we get into the math of this, let's think for a moment about current, voltage, and resistance and how they relate to each other.

Try this little quiz. Think in terms of the river if you find it helps.

1. If the voltage increases, will the current (a) increase or (b) decrease?
2. If the resistance increases, will the current (a) increase or (b) decrease?

Did you get the answers (a) and (b) correct, respectively?

If you write this down as an equation, it is called Ohm's law and can be written as:

$$I = V / R$$

I for current (I guess "C" was already taken), V for Volts, and R for resistance.

So, the current flowing through a resistor, or any wire connecting to it, will be the voltage across the resistor divided by the resistance of the resistor.

The units of resistance are in Ω (the abbreviation for ohms), while units of current are in A (short for amps, which is short for amperes) and in voltage V (the easy one).

So, if we have a voltage of 10V across a resistor of 100Ω the current flowing will be:

$$10V / 100\Omega = 0.1A$$

For convenience, we often use mA (1/1000 of an amp). So 0.1A is also 100mA.

That's enough about Ohm's law for now, we will meet it again later. It is the single most useful thing you can know about electronics. In the next section, we will look at the only other truly essential math you will need—power.

What Is Power?

Power is all about energy and time. So, in a way, it's a bit like current. But, instead of being the amount of charge passing a point, it is the amount of energy transformed into heat per second when a current passes through something that resists the flow (like a resistor). Forget the river, it doesn't really help much here.

Restricting the flow of a current generates heat, and the amount of heat generated can be calculated as the voltage across a resistor times the current flowing through it. The units of power are the watt (W). You would write this in math as:

$$P = I \times V$$

So, in our earlier example, we had 10V across a 100Ω resistor, so the current through the resistor was 100mA and will generate 0.1A × 10V, or 1 W of power. Given that the resistors that we have from the SparkFun kit are 250 mW (0.25 W). Our resistor will get hot and may eventually break.

If you don't know the current, but you do know the resistance, another useful formula for calculating the power is:

$$P = V^2 / R$$

Or, power is voltage squared (times itself) divided by the resistance. So, for the example earlier:

$$P = 10 \times 10 / 100 = 1\ W$$

That's reassuringly the same answer as we got before.

Most components have a maximum power rating like this, so when selecting a resistor, transistor, diode, and so on, it is worth doing a quick check and multiplying the voltage across the component by the current that you expect to flow through it. Then, choose a component with a maximum power rating comfortably greater than the expected power.

Power is the best measure of how much electricity is being used. It is the electrical energy being used per second, and unlike current it can be compared for devices operating from

Device	Power
Battery-powered FM radio (volume down)	20 mW
Battery-powered FM radio (volume up)	500 mW
Arduino Uno microcontroller board (9V supply)	200 mW
Home WiFi router	10 W
Compact fluorescent (low-power) light bulb	15 W
Filament light bulb	60 W
LCD TV 40-inch	200 W
Electric kettle	3000 W (3 kW)

TABLE 2-1 Power Usage

both 110 volt outlets and low voltage. It is good to have a basic understanding of just how much—or how little—electricity devices use. Table 2-1 shows some devices you might find around the home and lists how much power they use.

So, now you know why you don't get battery-powered kettles!

How to Read a Schematic Diagram

Hacking electronics often involves trawling the Internet, looking for people who have made something like the thing you want to make or adapt. You will often find schematic diagrams that tell you how to make and do things. So you need to be able to understand these schematics in order to turn them into real electronics.

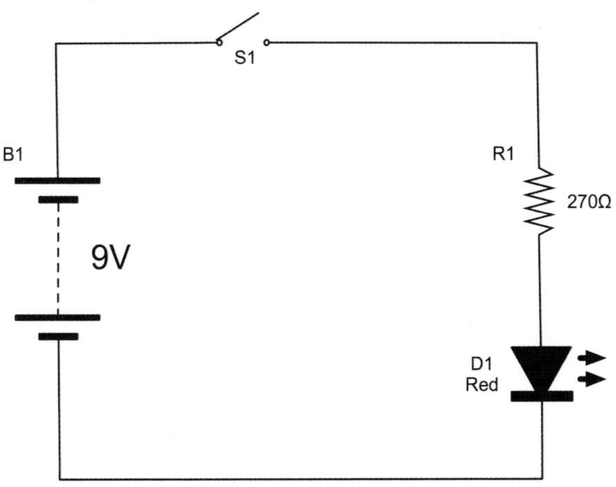

FIGURE 2-14 A simple schematic

These may at first sight seem a little baffling, but schematics obey a few simple rules and tend to use the same patterns over and over again. So there is a lot less to learn than you might think.

Ponder Figure 2-14 while we consider some of these rules—or more accurately conventions—because sometimes they are broken.

Figure 2-14 goes a long way to explaining why we sometimes talk of electronic circuits. It's kind of a loop. The current is flowing out of the battery, through the switch (when it's closed), through the resistor and LED (D1), and then back to the battery. The lines on the schematic can be thought of as perfect wires without any resistance.

The First Rule of Schematics: Positive Voltages Are Uppermost

A convention that most people follow when drawing a schematic is to put the higher voltages near the top, so on the left-hand side of the diagram, we have a 9V battery. The bottom of the battery is at 0V or GND (Ground), while the top of the battery will by 9V higher than that.

Notice that we draw the resistor R1 above the LED (D1). This way, we can think of some of the voltage as being lost across the resistor, before the remainder is lost through the diode and flows back to the negative connection of the battery.

Second Rule of Schematics: Things Happen Left to Right

Western civilization invented electronics and writes from left to right. You read from left to right and, culturally, more things happen from left to right. Electronics is no different, so it is common to start with the source of the electricity—the battery or power supply on the left—and then work our way from left to right across the diagram.

So, next we have our switch, which controls the flow of the electricity, and then the resistor and LED.

Names and Values

It is normal to give every component in a schematic a name. So, in this case the battery pack is called B1, the switch S1, the resistor R1, and the LED D1. This means that when you go from a schematic to a breadboard layout and eventually a circuit board, you can see which components on the schematic correspond to which components on the breadboard or circuit board.

It is also normal to specify the value of each of the components where appropriate. So, for example, the resistors' value of 270Ω is marked on the diagram. The rest of the components don't need much else said about them.

Component Symbols

Table 2-2 lists the most common circuit symbols you will encounter. This is nothing like a complete list, but we will discuss other symbols later in the book.

There are two main styles of circuit symbol: American and European. Fortunately, they are similar enough to avoid difficulties in recognizing them.

In this book, we will use the U.S. circuit symbols.

Symbol (U.S.)	Symbol (European)	Photo	Component	Use
	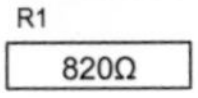 R1 820Ω	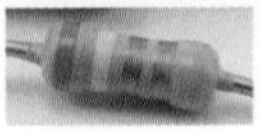	Resistor	Resisting
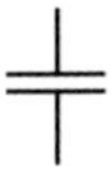	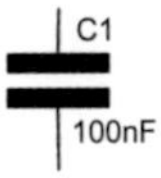 C1 100nF		Capacitor	Temporary charge storage
 +	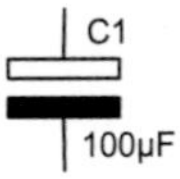 C1 100μF		Capacitor (polarized)	
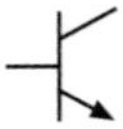			Transistor (bipolar NPN)	Using a small current to control a larger current
			Transistor (MOSFET N-channel)	Using a very small current to control a larger current
			Diode	Prevents current from flowing in the wrong direction
			LED	Indication and illumination
			Battery	Power supply
			Switch	Turning things on and off; control

TABLE 2-2 Common Schematic Symbols

Summary

In the next chapter, we get a much more practical look at some basic hacks and hone our electronic construction skills. This includes using prototyping boards and taking our soldering beyond simply connecting wires to other wires.

We will also learn how to use solderless breadboard so we can build electronics quickly and get underway.

AREF
GND
DIGITAL (PWM~)
UNO
ARDUINO
TX
RX
ON

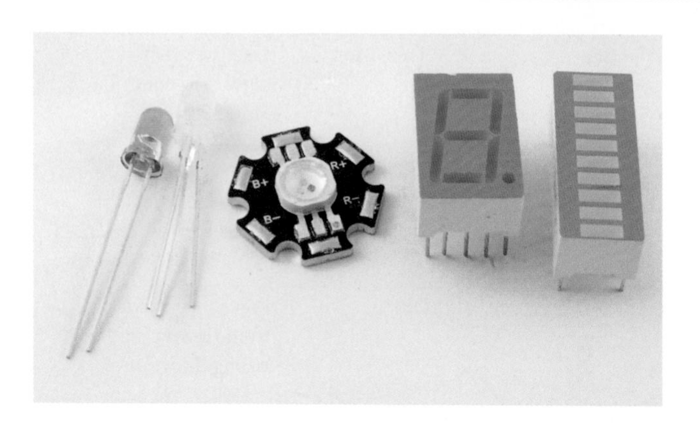
B+
R+
B–
R–

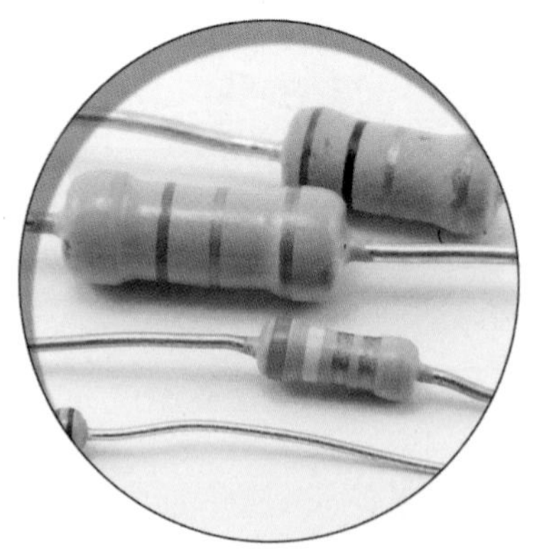

3

Basic Hacks

This chapter contains a set of fairly basic "hacking" how-tos. These build and use various electronic construction techniques. So this is probably a good chapter to at least skim through so that when you attempt more advanced how-tos, you can refer back to it if needed.

How to Make a Resistor Get Hot

Sometimes things will get hot when you are hacking electronics. It's always better when this is expected rather than when it's a surprise, so it's worth doing a little experimenting in this area.

You Will Need

Quantity	Item	Appendix Code
1	100Ω 0.25-watt resistor	K2
1	4 × AA battery holder	H1
1	4 × AA batteries (the rechargeable type is a good idea)	

Figure 3-1 shows the schematic diagram.

The Experiment

All we will do is connect the 100Ω resistor across the battery terminals and see how hot it gets.

Be careful when doing this because the resistor's temperature will rise to about 50°C/122°F. The resistor's leads, however, will not get very hot.

We are using a battery holder that takes four AA cells, each providing about 1.5V. They are each connected, one after the other, providing us with 6V total. Figure 3-2 shows how the batteries are actually connected within the battery box as a schematic diagram. In this kind of arrangement, the batteries are said to be in series.

Figure 3-3 shows the resistor heater in action.

Simply touch a finger to the resistor to confirm it's hot.

Is this bad/good? Will the resistor eventually break because it's warm? No, it won't. Resistors are designed to cope with a bit of heat. If we do the math, the power that the resistor is burning is the voltage squared divided by the resistance, which is:

$(6 \times 6) / 100 = 0.36W$

If it is a 0.25-W resistor, then we are exceeding its maximum power. This would be a foolish thing to do if we were designing a product for mass production. However, that's *not* what we are doing, and the chances are the resistor would continue to work like that indefinitely.

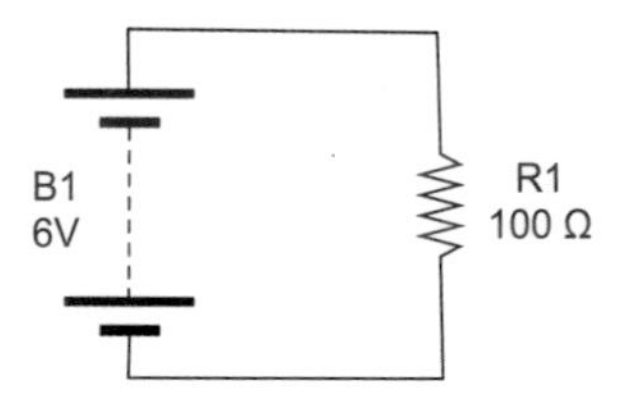

FIGURE 3-1 The schematic for heating a resistor

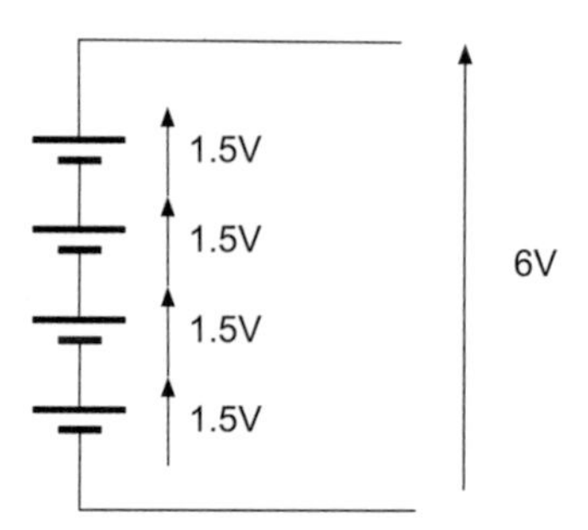

FIGURE 3-2 The schematic diagram for a battery holder

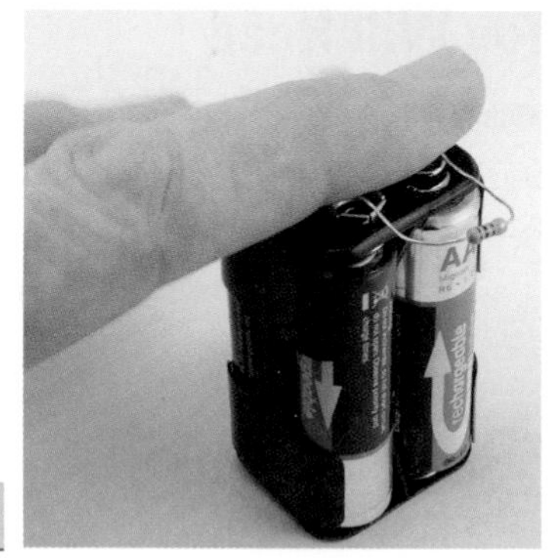

FIGURE 3-3 Making a resistor get hot

How to Use Resistors to Divide a Voltage

Sometimes voltages are too big. For example, in an FM radio, the signal going from the radio part to the audio amplifier part will be deliberately too large so it can be reduced using the volume knob.

Another example might be when you have a sensor that produces a voltage between 0 and 10V but you want to connect it to an Arduino microcontroller that expects it to be between 0 and 5V.

A very common technique in electronics is to use a pair of resistors (or a single variable resistor) as a "voltage divider."

You Will Need

Quantity	Item	Appendix Code
1	10kΩ trimpot (tiny variable resistor)	K1, R1
1	Solderless breadboard	T5
	Solid-core jumper wire	T6
1	4 × AA battery holder	H1
1	4 × AA batteries	
1	Battery clip	H2
1	Multimeter	T2

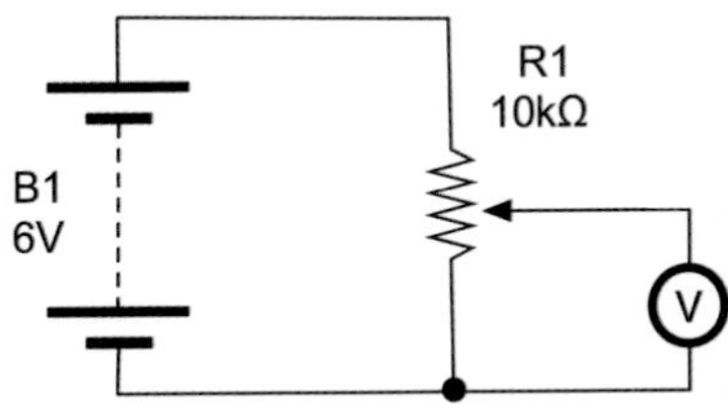

FIGURE 3-4 A voltage divider schematic diagram

Figure 3-4 shows the schematic diagram for our experiment. There are a couple of new schematic symbols here. The first is the variable resistor (or pot). This looks like a regular resistor symbol, but has a line with an arrow connecting to the resistor. This is the moving slider connection of the variable resistor.

The second new symbol is the circle with a "V" in it. This is a voltmeter, which in our case is the multimeter set to its DC voltage range.

The variable resistor we will use has three leads. One lead is fixed at each end of a conductive track, while a third connection to the central slider moves from one end of the track to the other. The overall resistance of the whole track is 10kΩ.

Our voltage in is going to be supplied from the battery pack and will be roughly 6V. We are going to use a multimeter to measure the output voltage and see how much it is being reduced by our voltage divider.

If you remember, the grey bars indicate where the connections underneath the holes are connected together. Take some time to follow the lines on the stripboard and reassure yourself that everything is connected in the same way as the schematic (Figure 3-4).

Plug the trimpot into the breadboard as shown, and then wire up the battery by carefully pushing the leads into the + and – power supply lines on the breadboard: red to +, black to –. If you struggle to get the multi-core wires of the battery clip into the holes, solder a bit of solid-core wire to the end of the leads.

Attach wires between the positive supply and the top connection of the trimpot, and the negative supply and the bottom connection of the trimpot. Finally, attach the multimeter. If your multimeter has alligator clips, use these in preference to the normal probes, clipping short jumper wires into the alligator clips and then pushing the other ends into the positions shown in Figure 3-5. When you have done all this, your breadboard should look something like Figures 3-6a and 3-6b.

Turn the trimpot to its fully clockwise position. The multimeter should read 0V (Figure 3-6a). Now turn it fully anti-clockwise and it should read something around 6V (Figure 3-6b)—in other words, the full battery voltage. Finally, turn it to roughly the middle position and you should see that the meter indicates about 3V (Figure 3-6c).

Think of the variable resistor as behaving a bit like two resistors, R1 and R2, as shown in Figure 3-7.

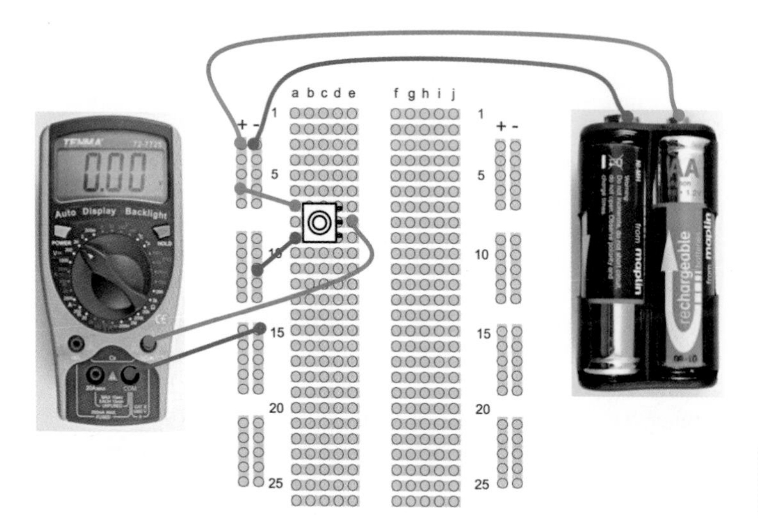

FIGURE 3-5 A voltage divider breadboard layout

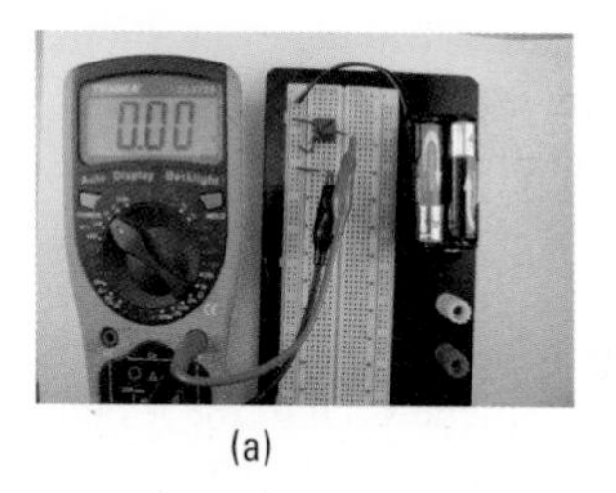

(a)

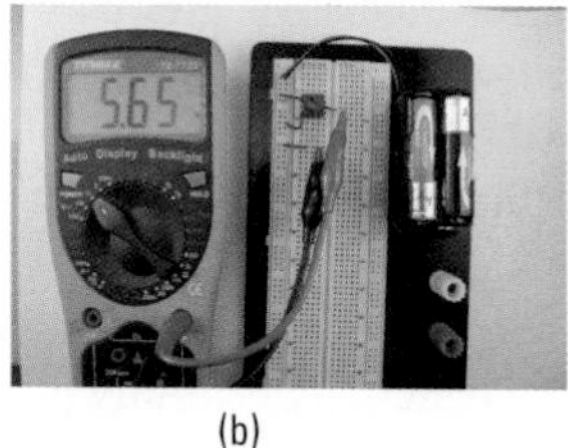

(b)

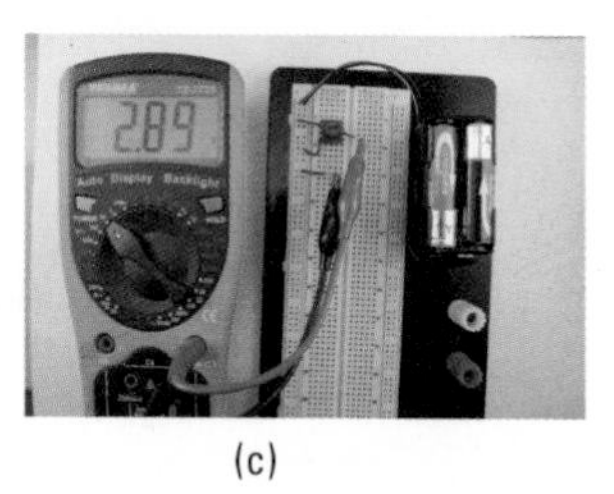

(c)

FIGURE 3-6 A voltage divider breadboard

The formula to calculate Vout if we know Vin, R1 and R2 is as follows:

Vout = Vin * R2 / (R1 + R2)

So, if R1 and R2 are both 5 kΩ and Vin is 6V, then:

Vout = 6V * 5kΩ / (5kΩ + 5kΩ) = 30 / 10 = 3V

This ties in with what we found when we put the trimpot to its middle position. It is exactly the same as having two fixed resistors of 5 kΩ each.

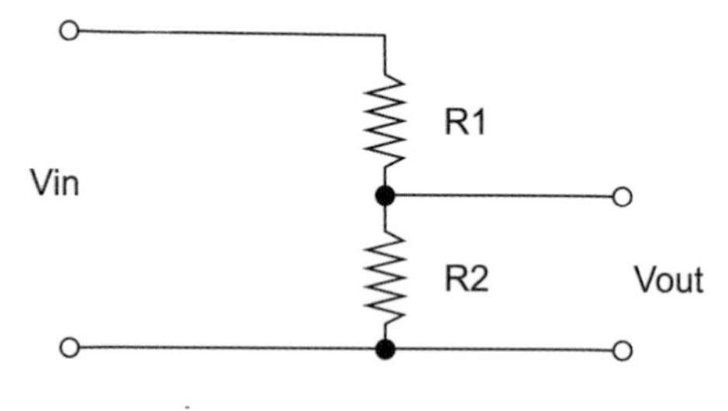

FIGURE 3-7 A voltage divider with fixed resistors

As with many of the calculations you make in electronics, people have made handy calculating tools. If you type "voltage divider calculator" into a search engine, you will find them. One such example can be found here: www.electronics2000.co.uk/calc/potential-divider-calculator.php.

These calculators will also usually match to the nearest available fixed resistor value.

How to Convert a Resistance to a Voltage (and Make a Light Meter)

An LDR (light-dependent resistor; a.k.a., photoresistor) is a resistor whose resistance changes depending on the amount of light falling on its transparent window. We will use one of these devices to demonstrate the idea of converting a resistance to a voltage by using it as one-half of a potential divider.

You Will Need

Quantity	Item	Appendix Code
1	Light-dependent resistor	K1, R2
1	Solderless breadboard	T5
	Solid-core jumper wire	T6
1	4 × AA battery holder	H1
1	4 × AA batteries	
1	Battery clip	H2
1	Multimeter	T2

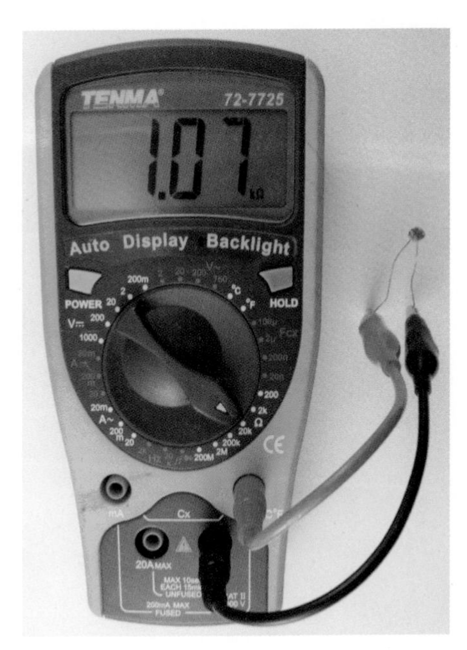

FIGURE 3-8 Measuring the LDR resistance

Before we get the breadboard out, let's just experiment directly with the LDR. Figure 3-8 shows the LDR connected directly to the multimeter on its 20kΩ resistance setting. As you can see, the resistance of my LDR was 1.07kΩ. Putting my hand over the LDR to screen out some of the light increased that resistance to a few tens of kΩ. So, the way the LDR works, the more light that reaches it, the lower the resistance.

Microcontrollers such as the Arduino can measure voltages and do things with them, but not directly measure resistance. So to convert our LDR's resistance into a more easily used voltage, we can put it in a voltage divider as one of the resistors (Figure 3-9).

Note that the symbol for the LDR is like a resistor but with little arrows pointing to it to indicate its sensitivity to light.

We can make up this schematic on our breadboard, this time setting our multimeter to the 20V DC range and watching how the voltage changes as we cover the LDR to reduce the light getting to it (Figures 3-10 and 3-11).

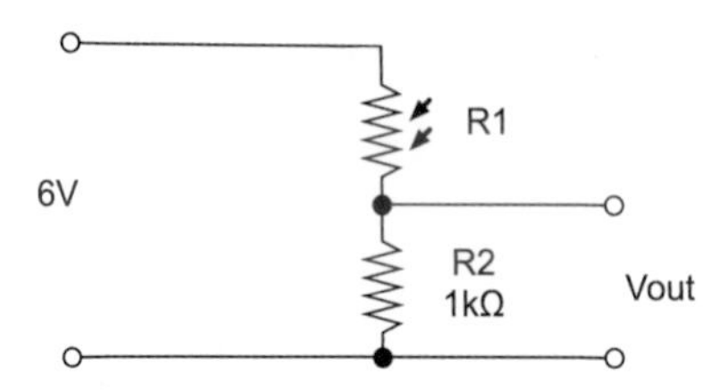

FIGURE 3-9 Measuring light level with an LDR and voltage divider

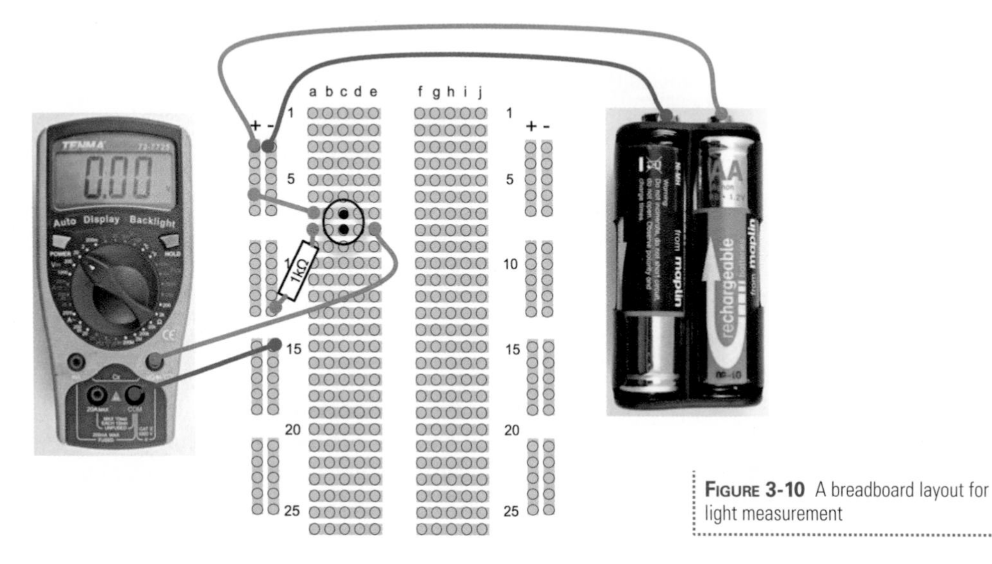

FIGURE 3-10 A breadboard layout for light measurement

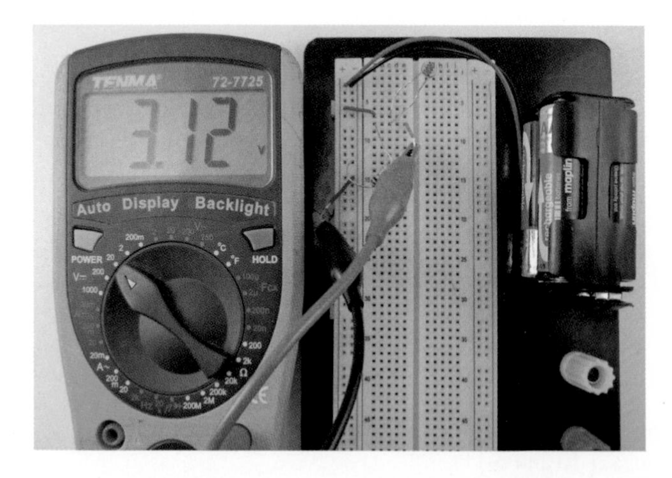

FIGURE 3-11 Light measurement

Hack a Push Light to Make It Light Sensing

Battery-powered push lights are one of the many glorious bargains you are likely to find in a dollar/euro/pound store. These are intended for use in cupboards and other dark locations where a bit of extra light would be useful. Push them once and they light, push them again and they turn off.

It will not surprise you to hear that we are going to use our LDR to turn the light on and off. But we are also going to use a transistor.

Our approach will be to get it working on breadboard first and then solder up the design onto the push light. In fact, we will use a single LED in place of the push lamp until we know that it will work.

You Will Need

Quantity	Name	Item	Appendix Code
1	R1	Light-dependent resistor	K1, R2
1	T1	Transistor 2N3904	K1, S1
1	R2	Resistor 10kΩ	K2
1*	R3	Resistor 220Ω	K2
1*	D1	Red LED or high-brightness LED	K1 or S2
*		Solid-core jumper wire	T6
1		Push light	

* These components are only needed for the breadboard experiment.

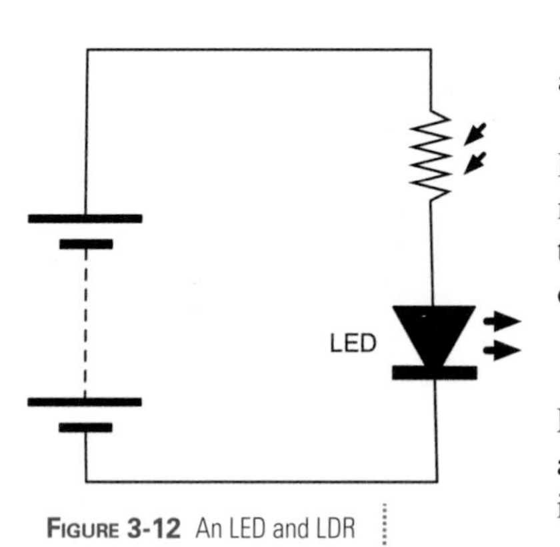

FIGURE 3-12 An LED and LDR

We want the LDR to control an LED, so a first thought at a circuit might be as shown in Figure 3-12.

There are two fatal flaws in this design. First, as more light falls on the LDR its resistance decreases, allowing more current to flow so the LED will get brighter. This is the opposite of what we want. We want the LED to come on when it's dark.

We need to use a transistor.

The basic operation of a transistor is shown in Figure 3-13. There are many different types of transistors, and probably the most common (and the type we will use) is called an NPN bipolar transistor.

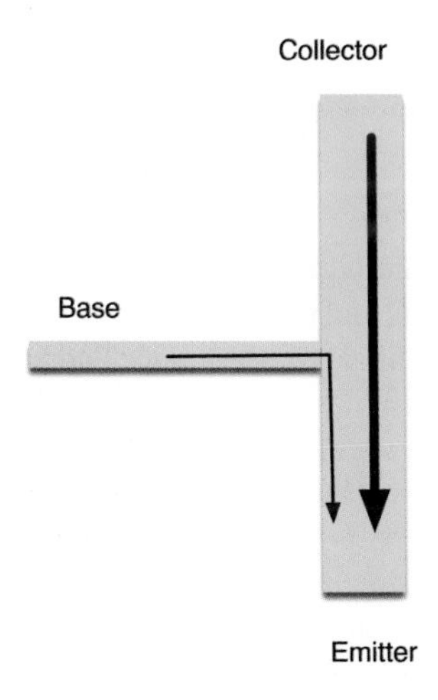

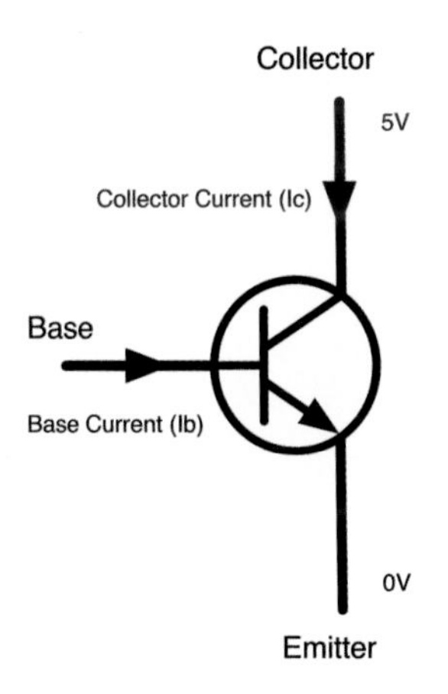

FIGURE 3-13 A bipolar transistor

This transistor has three leads: the emitter, the collector, and the base. The basic principal is that a small current flowing through the base will allow a much bigger current to flow between the collector and the emitter.

Just how much bigger the current is depends on the transistor, but it's typically a factor of 100.

Breadboard

Figure 3-14 shows the schematic diagram we will build on the breadboard. To understand this circuit, let's consider two cases.

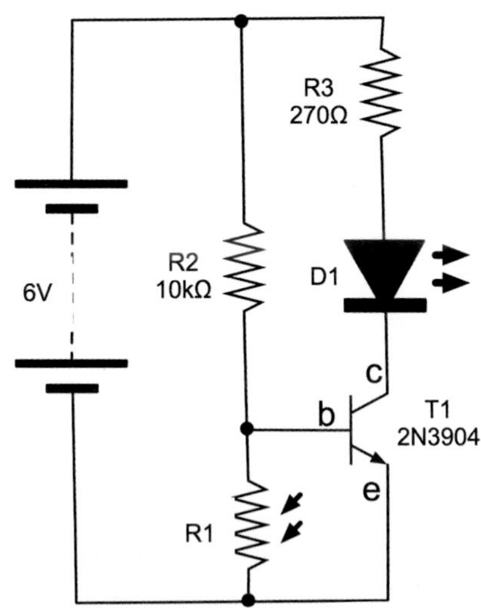

FIGURE 3-14 Using an LDR and transistor to switch an LED

Case 1: When It's Dark

In this case, the LDR R1 will have a very high resistance, so you could almost imagine that it isn't there at all. In that case, current will flow through R2, through the base and emitter of the transistor, allowing as much current as it needs to flow through R3, the LED, and T1 into its collector and out through the emitter. When enough current flows into the base of a transistor to allow current to flow from the collector to the emitter, this is called "turning on" the transistor.

We can calculate the base current using Ohm's law. In this situation, the base of the transistor will be at only about half a volt, so we can assume there is more or less the full 6V across the 10kΩ resistor R2. Since I = V / R, the current will be 6 / 10,000 A or 0.6mA.

Case 2: When It's Light

When it is light, we have to consider the resistance of the LDR R1. The lighter it is, the lower the resistance of R1 and the more of the current otherwise destined for the base of the transistor will be diverted through R1, preventing the transistor from turning on.

I think the time has come to build the project on breadboard. Figure 3-15 shows the breadboard layout, and Figures 3-16a and 3-16b the finished breadboard.

When placing the LED on the breadboard, make sure you get it the right way around. The longer lead is the positive lead, and it should be on row 10 connected to R3. (See Figure 3-16a.)

If everything is fine, you should find that when you cover the LDR, the LED should light (Figure 3-16b).

Construction

Now that we have proved our circuit works, we can get on with modding the push light. Figure 3-17 shows the push light the author used. Unless you are very lucky, yours is likely to be

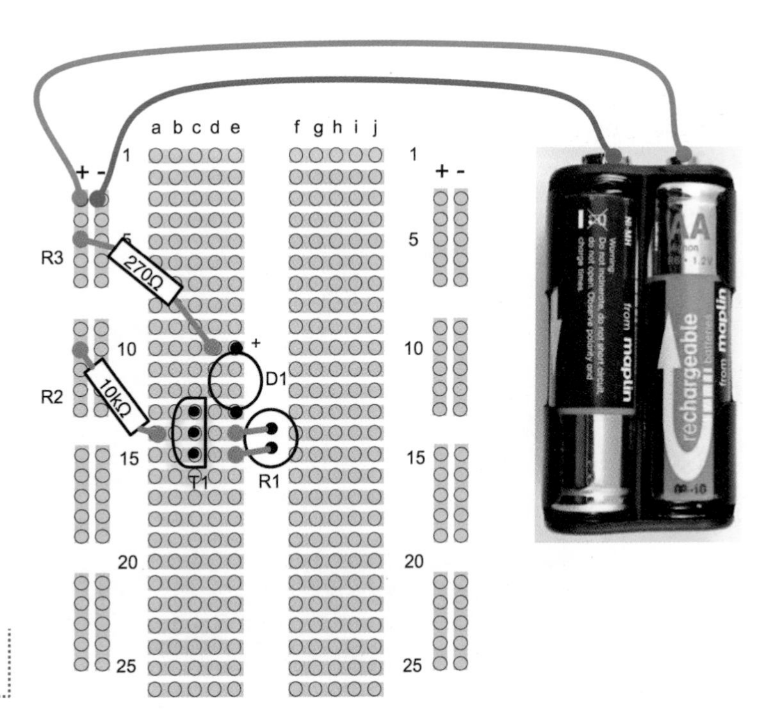

FIGURE 3-15 The light switch breadboard layout

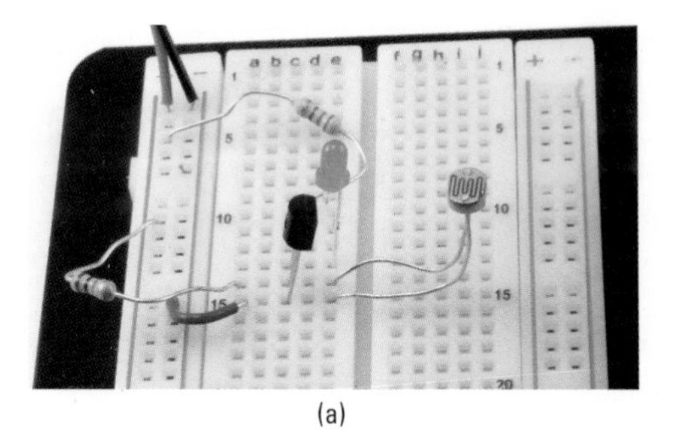
(a)

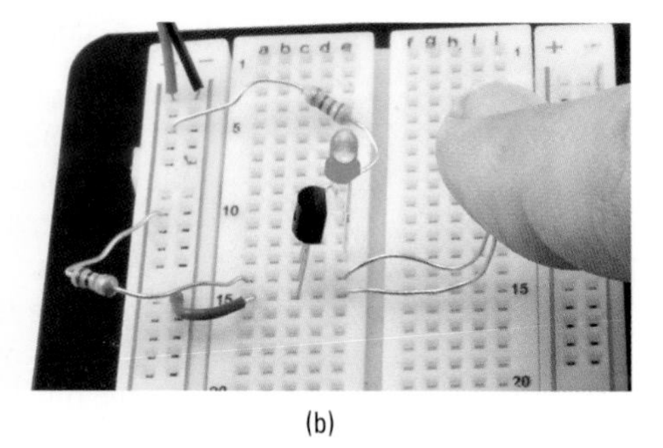
(b)

FIGURE 3-16 The light switch breadboard

different, so read through the following sections carefully and you should be able to work out how to change your light. To make life easy for yourself, try and find a push light that operates from 6V (4 AA or AAA cells).

You will probably find screws on the back of the push light. Remove these and put them somewhere safe. The inside of the push light is shown in Figure 3-18. The various connections on the light are marked. You can find the corresponding connections on your light using a multimeter.

Setting the multimeter to its 20V DC range will let you determine which battery lead is positive and which is negative. Looking at the wiring, we can draw a schematic diagram for the light as it stands, before we start altering it (Figure 3-19).

FIGURE 3-17 A push light

FIGURE 3-18 Inside the push light

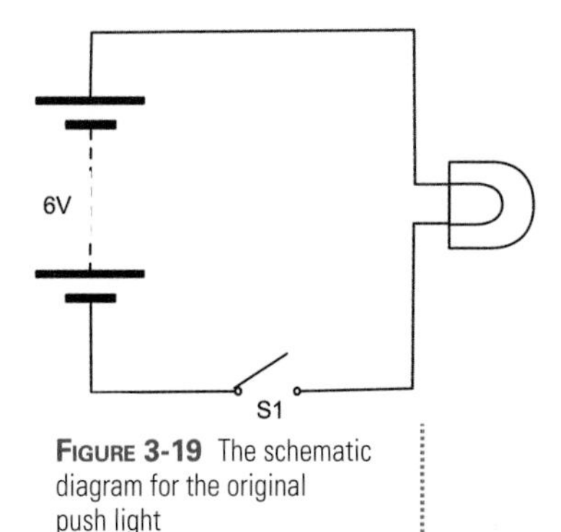

Figure 3-19 The schematic diagram for the original push light

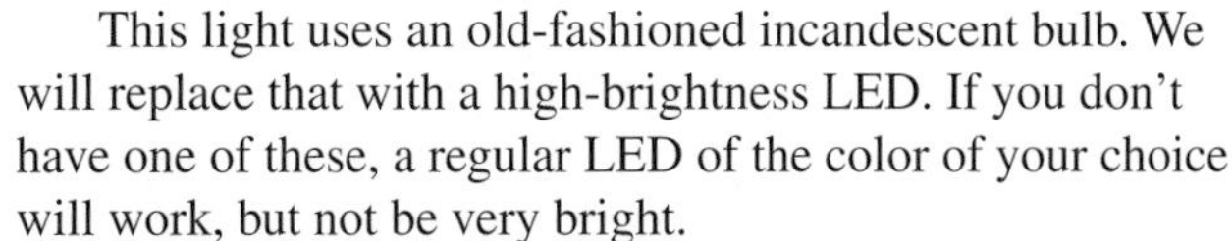

This light uses an old-fashioned incandescent bulb. We will replace that with a high-brightness LED. If you don't have one of these, a regular LED of the color of your choice will work, but not be very bright.

Figure 3-20 shows how we replaced the bulb with the LED and the 220Ω resistor. Make sure the longer positive lead of the LED is connected to the resistor and the far side of that resistor is connected to the positive terminal of the battery.

Try pressing the switch to make sure the LED is working.

We can now draw a schematic that combines what we have in the existing lamp and our LDR circuit (Figure 3-21).

In fact, all this really amounts to is adding in the switch to the original LED schematic. We have already installed R3 and D1 when we replaced the bulb with an LED. The switch is already there, so all we need to add is the transistor, LDR, and R2. Figure 3-22 shows how we will rewire the push light.

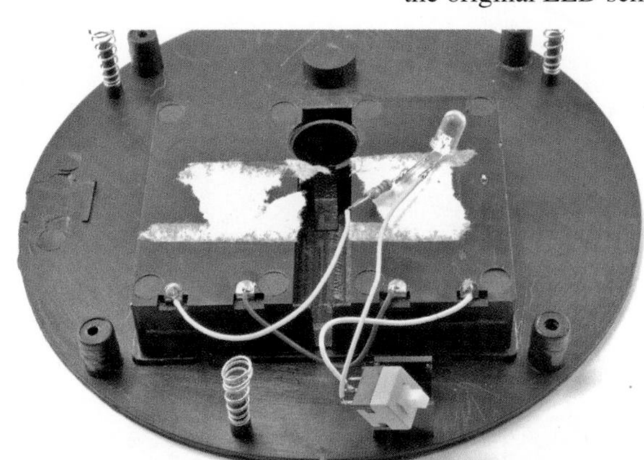

Figure 3-20 Replacing the bulb with an LED and a resistor

Figure 3-23 shows the sequence of steps in soldering the extra components onto the light.

1. Start by desoldering the lead from the switch that isn't connected to the negative battery terminal (Figure 3-23a).
2. Solder the 10kΩ resistor R2 between the middle lead of the transistor (the base) and the positive terminal on the battery box.
3. With the flat of the transistor facing upward, as shown in the diagram, connect the left-hand lead of the transistor to the wire you just disconnected from the switch (Figure 3-23b).
4. Solder the LDR between the left and middle pins of the transistor, and connect the combined left-hand transistor lead and LDR lead to the connection on the switch that the wire used to be attached to. (See Figure 3-23c.)
5. Tuck the components away neatly, bending the leads to make sure there is no way the bare leads can touch each other. (See Figure 3-23d.)

There you are! You have hacked some electronics.

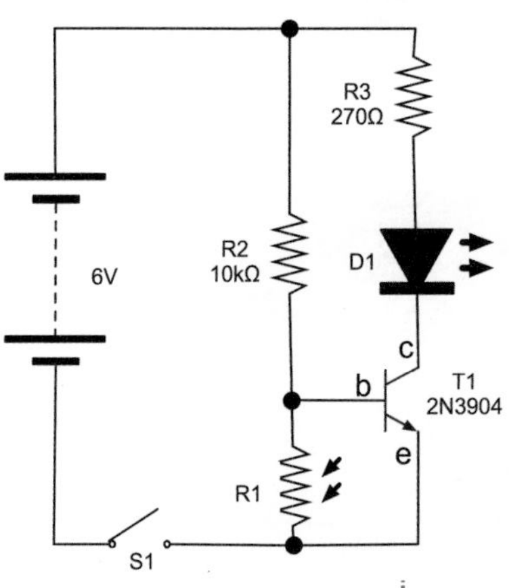

Figure 3-21 The final schematic

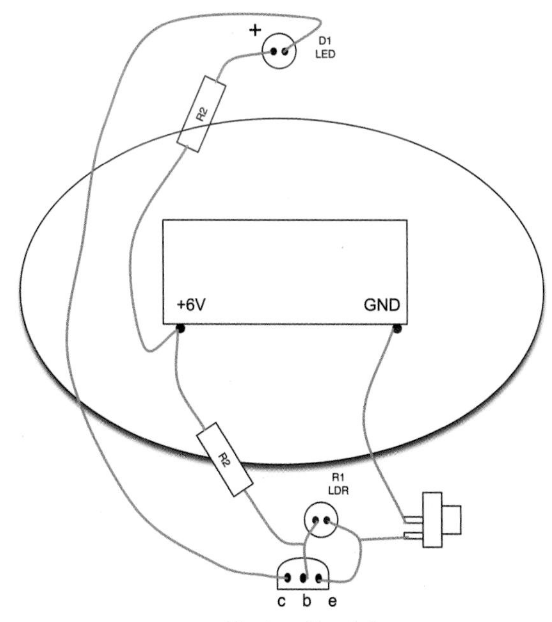

FIGURE 3-22 The push light wiring diagram

FIGURE 3-23 Soldering the project

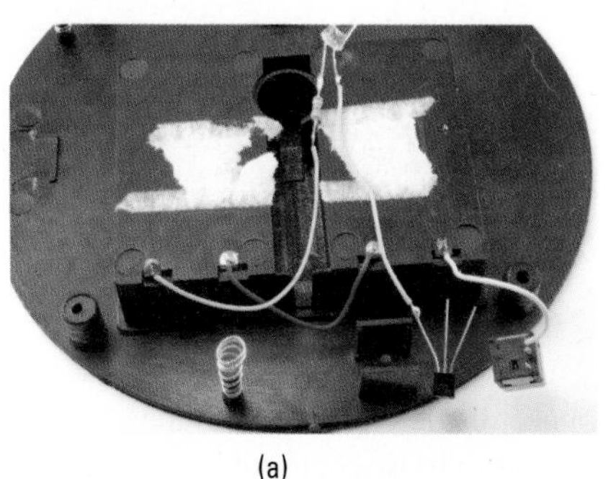

(a)

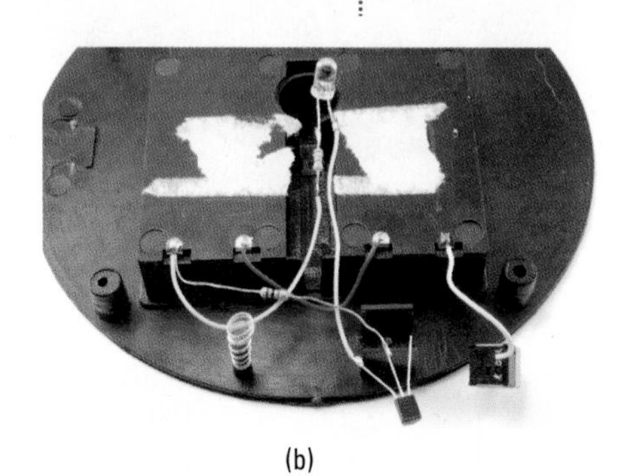

(b)

(c)

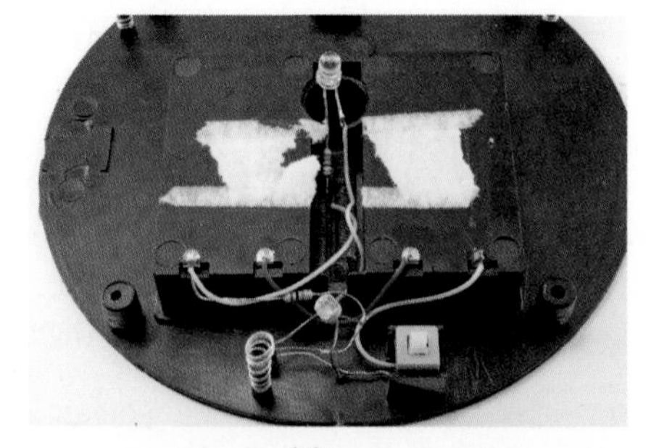

(d)

How to Choose a Bipolar Transistor

The transistor we used in the previous section, "Hack a Push Light to Make It Light Sensing," is a useful general-purpose transistor. But there are many other types of transistors that we can use for different purposes. This section is to help you find the right transistor and try to use it in such a way that it doesn't die in a puff of smoke.

Datasheets

Transistors have a number of parameters we need to know about. All transistors have an associated datasheet. This is produced by the manufacturer and specifies everything you could possibly want to know about the device, from the dimensions of its leads to its electrical characteristics.

Most of the time you will use one of three or four transistors and will not need to look into the exact details of how the transistor behaves, but if you do need to, it's there on the datasheet. So, you may just want to skip to the next subsection where we look at a few different types of transistors—just the useful ones, nothing exotic.

Table 3-1 shows some of the data you will find on the 2N3904's datasheet under maximum ratings.

Maximum Collector-Emitter and Collector-Base voltages of 40V and 60V mean we do not have to worry about exceeding them in battery-powered devices. We need to be careful that we do not exceed the emitter-base voltage though.

The maximum collector current of 200mA is quite healthy though. It means we could in theory control ten LEDs, all taking 20mA at the same time. If we do exceed this value, then the transistor will get hot and eventually fail.

The one electrical characteristic we are most interested in is DC current gain or h_{FE} as it will be called on the datasheet. This is listed in the electrical characteristics section of the datasheet.

Absolute Maximum Ratings			
Symbol	Parameter	Value	Units
V_{CEO}	Collector-Emitter Voltage	40	V
V_{CBO}	Collector-Base Voltage	60	V
V_{EBO}	Emitter-Base Voltage	6.0	V
I_C	Collector Current – Continuous	200	mA

TABLE 3-1 2N3904 Maximum Ratings

You may remember that the DC gain is the multiplier that determines how much more current can flow in through the base than the collector. Looking at Table 3-2, this means that at a collector current of 10mA and a collector emitter voltage of 1.0V (it's

Symbol	Parameter	Test Condition	Min	Max	Units
ON CHARACTERISTICS					
h_{FE}	DC Current Gain	$I_C = 0.1$ mA, $V_{CE} = 1.0$V	40		
		$I_C = 1.0$ mA, $V_{CE} = 1.0$V	70		
		$I_C = 10$ mA, $V_{CE} = 1.0$V	100	300	
		$I_C = 50$ mA, $V_{CE} = 1.0$V	60		
		$I_C = 100$ mA, $V_{CE} = 1.0$V	30		

TABLE 3-2 2N3904 Electrical Characteristics

nearly always about that), the typical gain will be 100, meaning that only 10mA / 100 = 100nA needs to be flowing into the base for this amount of current to flow through the collector.

MOSFET Transistors

The 2N3904 is what is called a bipolar transistor. It's basically a device that amplifies current. A small current into the base controls a much bigger current flowing through the collector. Sometimes, the current gain of just 100 or so is not nearly enough.

There is another type of transistor that does not suffer from this limitation called the MOSFET (Metal Oxide Semiconductor Field Effect Transistor). You can see why it gets shortened to MOSFET. These transistors are controlled by voltage rather than current and make very good switches.

MOSFETs do not have emitters, bases, and collectors, they have "sources," "gates," and "drains." They turn on when the gate voltage passes a threshold, usually about 2V. Once on, quite large currents can flow through the "drain" to the "source" rather like a bipolar transistor. But since the gate is isolated from the rest of the transistor by a layer of insulating glass, hardly any current flows into the gate. It is the voltage at the gate that determines what current will flow.

We will meet MOSFETs again later in the section "How to Use a Power MOSFET to Control a Motor," and in Chapter 7 in the section "How to Control Motor Speed with a Power MOSFET."

PNP and N-Channel Transistors

The automated light switch of the previous section switched on the "negative side of the load." That is, if you go back to Figure 3-21, you can see that the resistor and LED that make up the light are not connected to GND except through the

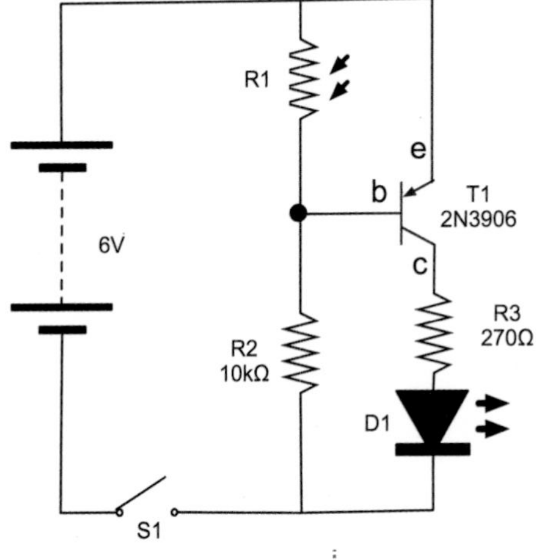

FIGURE 3-24 Using a PNP bipolar transistor

transistor. If for some reason (and this does happen) we wanted to switch the positive side, then we would need to use a PNP equivalent of the NPN 2N3904, such as the 2N3906. NPN stands for Negative-Positive-Negative, and yes, you can guess what PNP stands for. That is because transistors are kind of semiconductor sandwiches, with material of either N or P type as the bread. If the bread is N type (the most common), then the base voltage needs to be higher than the emitter voltage (by about 0.5V) before the transistor starts to turn on. On the other hand, a PNP transistor turns on when the base voltage is more than 0.5V lower than the emitter voltage.

If we wanted to switch the positive side, we could use a PNP transistor (as shown in the PNP alternative to Figure 3-21) displayed in Figure 3-24.

MOSFETs also have their own equivalent of PNP transistors called P-channel, their version of the more common NPN being called N-channel.

Common Transistors

The transistors in Table 3-3 will cover a wide range of transistor applications. There are thousands and thousands of other transistors, but in this book we only really use them for switching, so these will cover most "bases"!

Name	Appendix Code	Type	Max Switching Current	Notes
Low/medium-current switching				
2N3904	S1	NPN bipolar	200mA	Current gain about 100
2N3906	S4	PNP bipolar	200mA	Current gain about 100
2N7000	S3	N-channel MOSFET	200mA	2.1V gate-source threshold voltage; turns on when gate is 2.1V higher than source
High-current switching				
FQP30N06	S6	N-channel MOSFET	30A	2.0V gate-source threshold voltage; turns on when gate is 2.0V higher than source

TABLE 3-3 Really Useful Transistors

How to Use a Power MOSFET to Control a Motor

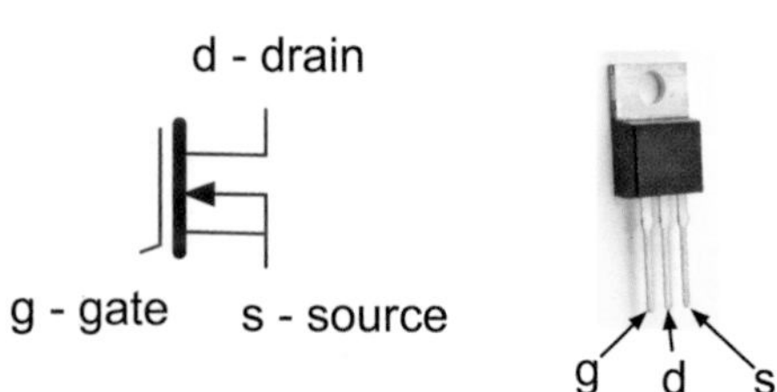

FIGURE 3-25 The FQP30N06 N-channel MOSFET

Figure 3-25 shows the schematic symbol and the pinout for the FQP30N06 N-channel MOSFET.

This MOSFET is capable of controlling loads of up to 30A. We are not going to push it any way near that far, we are just going to use it to control the power to a small electric motor that might have a peak load of 1 or 2A. While this would be too much for the bipolar transistors that we have been using so far, this MOSFET will hardly notice!

You Will Need

To try out this high-power MOSFET, you will need the following items.

Quantity	Item	Appendix Code
1	Solderless breadboard	T5
	Solid-core jumper wire	T6
1	4 × AA battery holder	H1
1	4 × AA batteries	
1	Battery clip	H2
1	Multimeter	T2
1	10kΩ trimpot	K1
1	FQP30N06 MOSFET	S6
1	6V DC motor or gear motor	H6

The DC motor can be any small motor you can find that is around 6V. A motor rated at 12V should still turn at 6V. To test it, just connect its terminals directly to the 6V battery.

Breadboard

The schematic diagram for what we will make is shown in Figure 3-26.

The variable resistor will control the voltage at the gate of the MOSFET. When that gate voltage exceeds the gate threshold, the transistor will turn on and the motor will start.

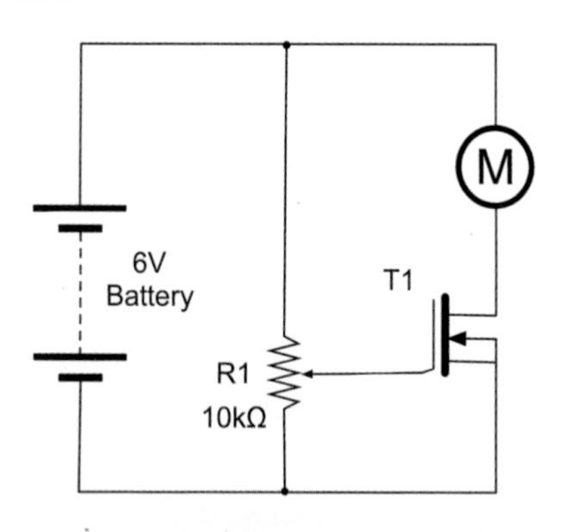

FIGURE 3-26 A schematic for the MOSFET experiment

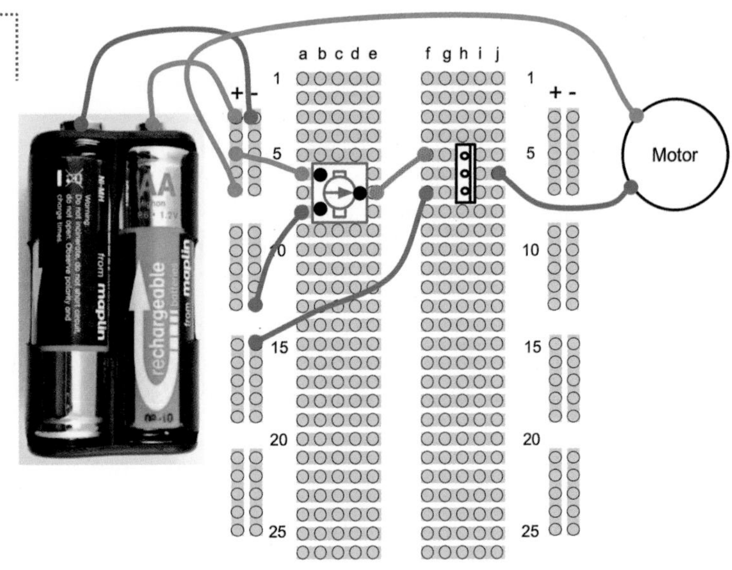

FIGURE 3-27 The breadboard layout for the MOSFET experiment

The breadboard layout for the project and a photograph of the experiment in action are shown in Figure 3-27 and Figure 3-28.

To connect the motor to the breadboard, you will probably need to solder a pair of leads to it. It does not matter which way around you connect the motor. The polarity just determines which direction the motor turns. So if you swap the motor leads over, it will turn in the opposite direction.

Try turning the knob on the variable resistor. You will see that you do not have a great deal of control over the speed of

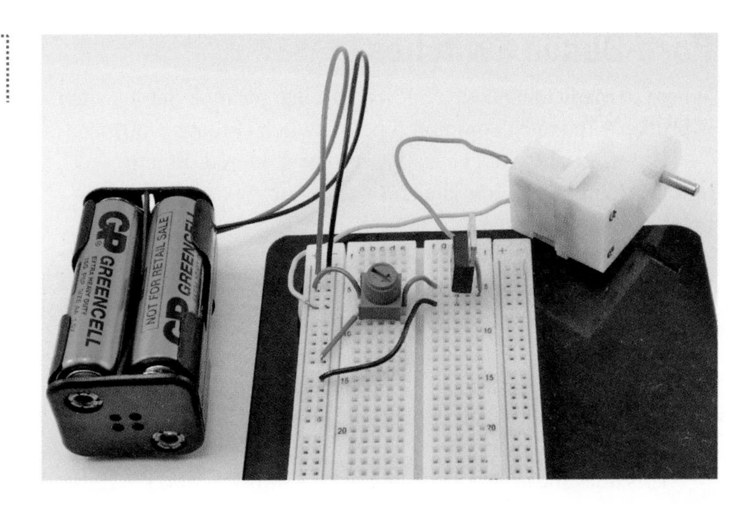

FIGURE 3-28 The MOSFET experiment

the motor. If you hover around the threshold voltage, you can control the motor speed, but you can probably see why the MOSFET is most commonly used as a switch that is either on or off.

This kind of MOSFET is called a logic level MOSFET, because its gate voltage is low enough to be controlled directly by digital output pins on a microcontroller. This is not true of all MOSFETs. Some have gate threshold voltages of 6V or more.

In Chapter 7, you will use a MOSFET to finely control the motor speed.

How to Select the Right Switch

On the face of it, a switch is a very simple thing. It closes two contacts, making a connection. Often, that is all you need, but other times you will require something more complicated. For example, let's say you want to switch two things at the same time.

There are also switches that only make the contact while you are pressing them, or ones that latch in one position. Switches may be push button, toggle, or rotary. There are many options to choose from and in this section we will attempt to explain the options.

Figure 3-29 shows a selection of switches.

FIGURE 3-29 Switches

Push-Button Switches

Where so many things use a microcontroller, a simple push switch is probably the most common type of switch (Figure 3-30).

This kind of switch is designed to be soldered directly onto a circuit board. It will also fit onto our breadboard, which makes it quite handy.

The confusing thing about this switch is that it has four connections where you would only expect there to be two. Looking at Figure 3-30, you can see that connections B and C are always connected together, as are A and D. However, when the button is pressed, all four pins are all connected together.

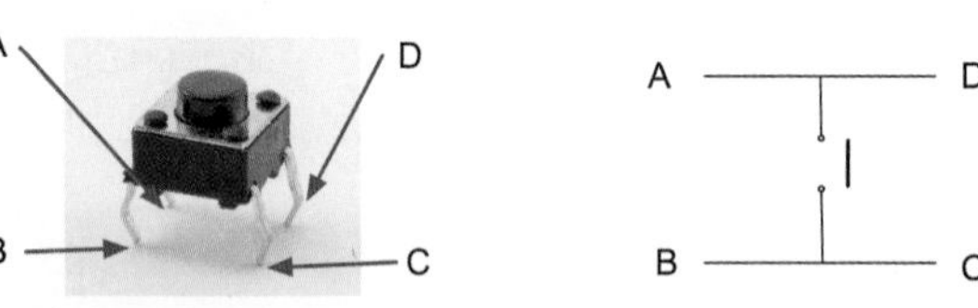

FIGURE 3-30 A push switch

This does mean that you need to be careful to find the right pins or your switch will be connected all the time.

If there is any doubt about how the switch works, use your multimeter set to Continuity mode to work out what is connected to what—first without the switch pressed and then with the switch pressed.

Microswitches

A microswitch is another type of handy switch. They are not designed to be pressed directly, but are often used in things like microwave ovens to detect that the door is closed, or as anti-tamper switches that detect when the cover is removed from an intruder alarm box.

Figure 3-31 shows a microswitch—with three pins!

The reason a microswitch has three pins rather than just two is that it is what is known as a "double throw" or "change-over" switch. In other words, there is one common connection C and two other connections. The common connection will always be connected to one of those contacts, but never both at the same time. The normally open (n.o.) connection is only closed when the button is pressed; however, the normally closed (n.c.) connection is normally closed, and only opens when the button is released.

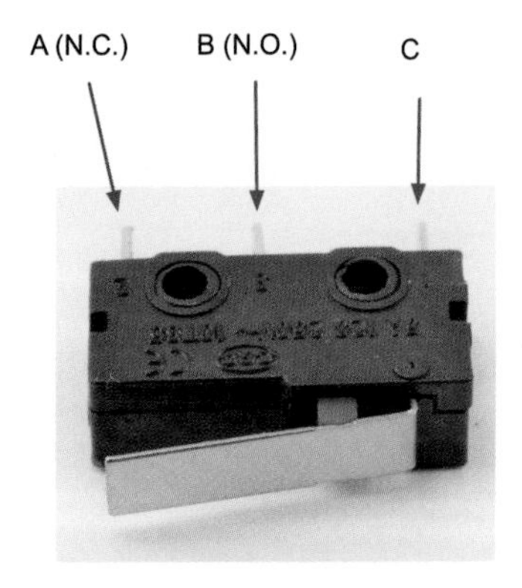

FIGURE 3-31 A microswitch

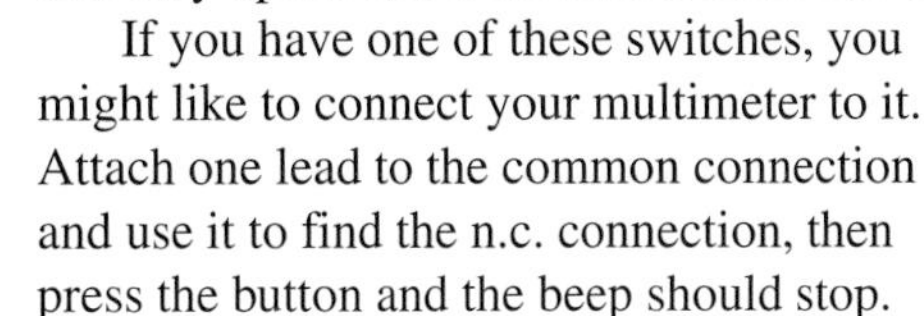

If you have one of these switches, you might like to connect your multimeter to it. Attach one lead to the common connection and use it to find the n.c. connection, then press the button and the beep should stop.

Toggle Switches

If you look through a component catalog (which every good electronics hacker should), you will find a bewildering array of toggle switches. Some will be described as DPDT, SPDT, SPST, or SPST, momentary on, and so forth.

Let's untangle some of this jargon, with a key for these cryptic letters:

- D = Double
- S = Single
- P = Pole
- T = Throw

So, a DPDT switch is double pole, double throw. The word "pole" refers to the number of separate switch contacts that are controlled from the one mechanical lever. So, a double pole switch can switch two things on and off independently. A single throw switch can only open or close a contact (or two contacts if it is double pole). However, a double throw switch can make the common contact be connected to one of two other contacts. So, a microswitch is a double throw switch because it has both normally closed and normally open contacts.

Figure 3-32 summarizes this.

Notice in Figure 3-32 that when drawing a schematic with a double pole switch, it is normal to draw the switch as two switches (S1a and S1b) and connect them with a dotted line to show they are linked mechanically.

The matter is further complicated because you can have three poles or even more on a switch, and double throw switches are sometimes sprung, so they do not stay in one or both of these positions. They may also have a center-off position where the common contact is not connected to anything.

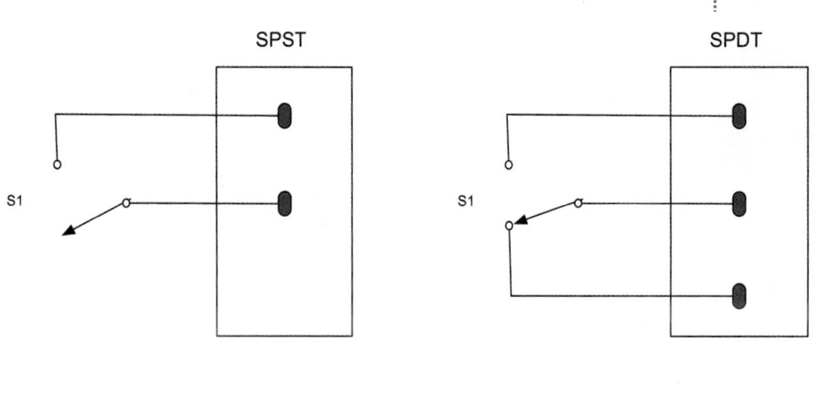

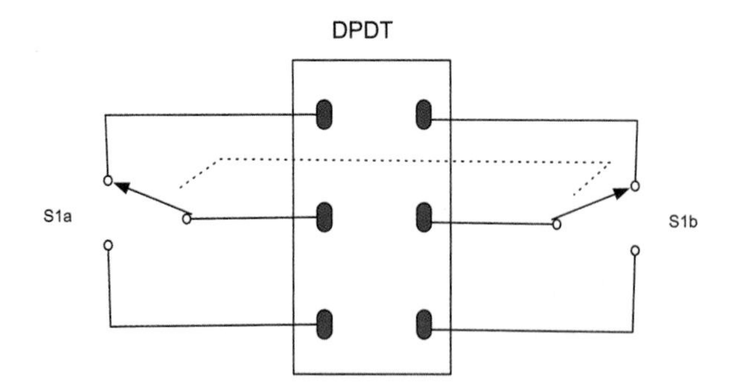

Figure 3-32 Toggle switches—poles and throws

You might see a switch described as "DPDT, On-Off-Mom." Well, we know what the DPDT bit means. It will have six legs for a start. The "On-Off-Mom" part means that it also has a center position, where the common connection is not made to anything. Switch it one way and it will be on to one set of contacts and stay in that position. Switch it the other way and it will be sprung to return to the central position, allowing you to make a "momentary" connection.

A lot of this terminology applies to other kinds of switches in addition to toggle switches.

Summary

We now know a bit about voltage, current resistance, and power. In the next chapter, we will use these ideas in looking at how to use LEDs.

ARDUINO
UNO
DIGITAL (PWM~)
AREF
GND
TX
RX
ON

ON
1 2 3 4

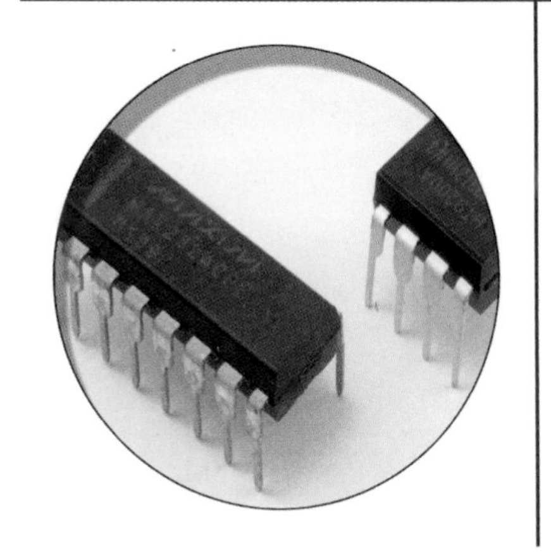

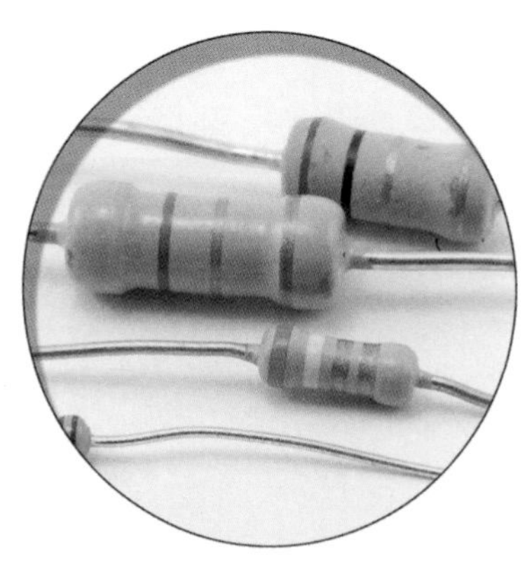

4

LEDs

LEDs (light-emitting diodes) are diodes that emit light when a current passes through them. Well on the way to completely replacing filament light bulbs in almost all applications, they can be used as indicators and, with the very high brightness types of LED, can provide illumination.

They are much more efficient than conventional light bulbs, producing far more light per watt of power, and are a lot less delicate.

LEDs do, however, require a little more thought when used. They have to be powered with the correct polarity and require circuitry to limit the current flowing through them.

How to Stop an LED from Burning Out

LEDs are delicate little things and quite easy to destroy accidentally. One of the quickest ways of destroying an LED is to attach it to a battery without using a resistor to limit the current.

To get to grips with LEDs, we will put three different color LEDs on our breadboard (Figure 4-1).

You Will Need

Quantity	Names	Item	Appendix Code
1		Solderless breadboard	T5
1	D1	Red LED	K1
1	D2	Yellow LED	K1
1	D3	Green LED	K1
1	R1	330Ω resistor	K2
2	R2, R3	220Ω resistor	K2
		Jumper wires	T6
1		4 × AA battery holder	H1
1		Battery clip	H2
4		AA batteries	

Diodes

We need to understand LEDs a little better if we are to successfully use them. LED stands for light-emitting diode, so let's start by looking at what a diode is (Figure 4-2).

A diode is a component that only lets current flow in one direction. It has two leads, one called the anode, and the other called the cathode. If the anode is at a higher voltage than the cathode (it has to be greater by about half a volt), then it will conduct electricity and is said to be "forward-biased." If on the other hand the anode isn't at least half a volt higher than the cathode, it is said to be "reverse-biased" and no current flows.

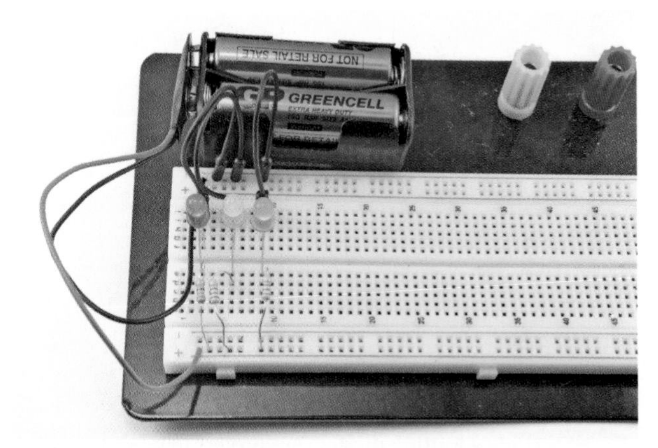

FIGURE 4-1 LEDs on a breadboard

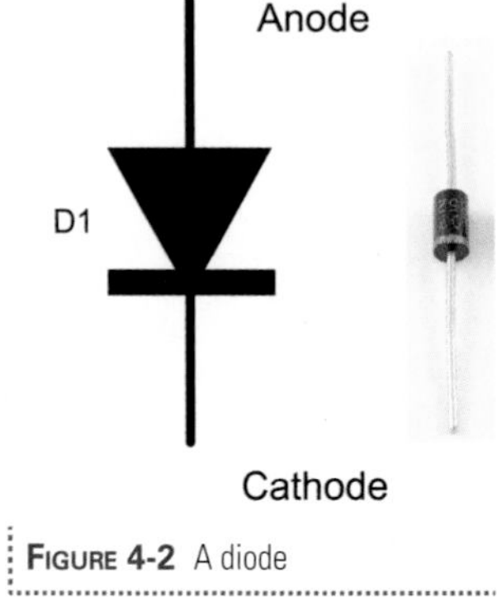

FIGURE 4-2 A diode

LEDs

An LED is just like a regular diode except that when it is forward-biased, it conducts and generates light. It also differs from a regular diode in that the anode usually needs to be at least 2V higher than the cathode for it to be forward-biased.

Figure 4-3 shows the schematic diagram for driving an LED.

The key to this circuit is to use a resistor to limit the current flowing through the LED. A normal red LED will typically just be lit at about 5mA and is designed to be used at around 10 to 20mA (this is called the "forward current" or I_F). We will aim for 15mA for our LED. We can also assume that when it is conducting, there will be about 2V across it. This is called the "forward voltage" or V_F. That means there will be $6 - 2 = 4V$ across the resistor.

So, we have a resistor that has a current flowing through it (and the LED) of 15mA and a voltage across it of 4V. We can use Ohm's law to calculate the value of resistance we need to achieve this:

$$R = V / I = 4V / 0.015A = 267\Omega$$

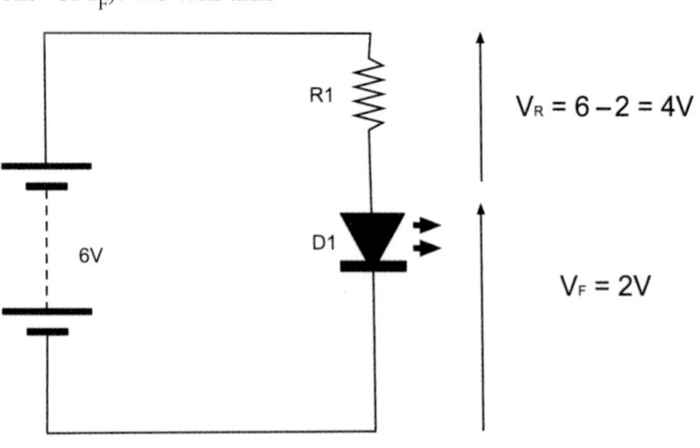

FIGURE 4-3 Limiting current to an LED

Parameter	Red	Green	Yellow	Orange	Blue	Units
Maximum forward current (I_F)	25	25	25	25	30	mA
Typical forward voltage (V_F)	1.7	2.1	2.1	2.1	3.6	V
Maximum forward voltage	2	3	3	3	4	V
Maximum reverse voltage	3	5	5	5	5	V

TABLE 4-1 An LED Datasheet

Resistors come in standard values, and the nearest higher value in our resistor starter kit is 330Ω.

As I mentioned earlier, a red LED will almost always light quite brightly with something like 10–20mA. The exact current is not critical. It needs to be high enough to make the LED light, but not exceed the maximum forward current of the LED (for a small red LED, typically 25mA).

Table 4-1 shows a section of the datasheet for a typical range of LEDs of different colors. Note how V_F changes for different color LEDs. This will mean you may need to use a different resistor, but usually if the supply voltage is above say 6V, then small variations in V_F for color will not require a different resistor value.

The other parameter you should be aware of is the "maximum reverse voltage." If you exceed this by, say, wiring your LED the wrong way around, it is likely to break the LED.

Many online series resistor calculators are available that—given the supply voltage V_F and current I_F for your LED—will calculate the series resistor for you. For example:

www.electronics2000.co.uk/calc/led-series-resistor-calculator.php

Table 4-2 is a useful rough guide, assuming a forward current of around 15mA.

Supply Voltage (V)	Red	Green, Yellow, Orange	Blue
3	91Ω	60Ω	none
5	220Ω	180Ω	91Ω
6	270Ω / 330Ω	220Ω	180Ω
9	470Ω	470Ω	360Ω
12	680Ω	660Ω	560Ω

TABLE 4-2 Series Resistors for LEDs

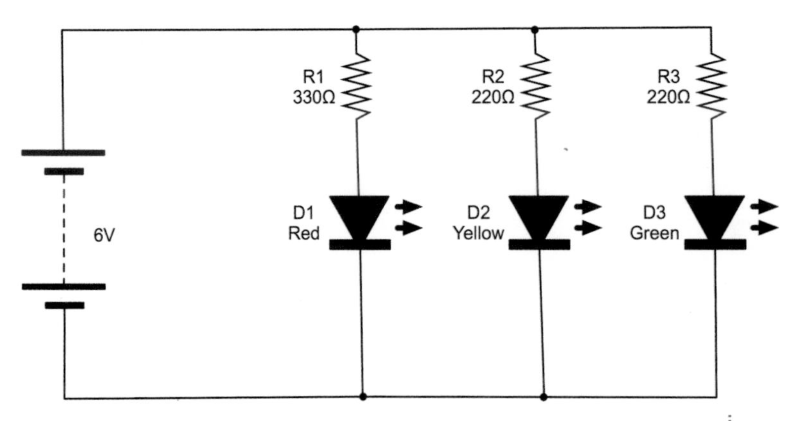

FIGURE 4-4 An LED's schematic

Trying It Out

You might like to try out your LEDs and get them lit up on the breadboard. So, using Figures 4-4 and 4-5 as a guide, wire up your breadboard. Remember that the longer lead of the LED is normally the anode (positive) and thus should be to the left of the breadboard.

An important point to notice here is that each LED has its own series resistor. It is tempting to use one lower value current limiting resistor and put the LEDs in parallel, but don't do this. If you do, the LED with the lowest V_F will hog all the current and probably burn out, at which point the LED with the next lowest V_F will do the same, until all the LEDs are dead.

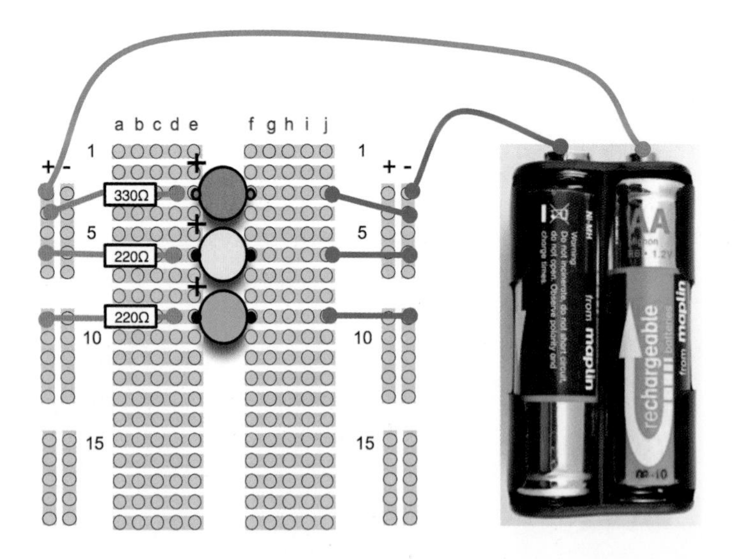

FIGURE 4-5 An LED breadboard layout

How to Select the Right LED for the Job

LEDs come in all colors, shapes, and sizes. Many times, you just want a little indicator light, in which case a standard red LED is usually fine. However, there are many other options, including LEDs bright enough to be used as lamps.

You Will Need

Quantity	Names	Item	Appendix Code
1		Solderless breadboard	T5
1	D1	RGB common cathode LED	S4
3	R1–R3	500Ω trimpot	R3
1	R1	330Ω resistor	K2
2	R2, R3	220Ω resistor	K2
		Jumper wires	T6
1		4 × AA battery holder	H1
1		Battery clip	H2
4		AA batteries	

Brightness and Angle

When selecting an LED, they may simply be described as "standard" or "high brightness" or "ultra-bright." These terms are subjective and open to abuse by unscrupulous vendors. What you really want to know is the LED's luminous intensity, which is how much light the LED produces. You also want to know the angle over which the LED spreads the light.

So, for a flashlight, you would use LEDs with a high luminous intensity and a narrow angle. Whereas for an indicator light to show that your gadget is turned on, you would probably use an LED with a lower luminous intensity but a wider angle.

Luminous intensity is measured in millicandela or mcd, and a standard indicator type LED will typically be around 10 to 100 mcd, with a fairly wide viewing angle being 50 degrees. A "high brightness" LED might be up to 2000 or 3000 mcd, and an ultra-bright anything up to 20,000 mcd. A narrow beam LED is about 20 degrees.

Multicolor

We have already explored the more common LED colors, but you can also get LED packages that actually contain two or three LEDs of different colors in the same package. Common varieties are red/green as well as full-color RGB (Red Green Blue). By varying the proportion of each color, you can change the color of the light produced by the LED package.

Figure 4-6 shows the schematic we can use to try out a little experiment with an RGB LED.

We are going to use a variable resistor with each of the red, green, and blue LEDs. The fixed resistors (R4, R5, and R6) are to prevent too much current flowing when the slider of the variable resistor is right at 6V.

Figure 4-7 shows the breadboard layout for this. The common lead of the LED is the longest lead, while the other three are the three-color anodes.

Once all the components are in the board and you have attached the battery, you should be able to mix various colors by changing the position of the three sliders. Figure 4-8 shows the circuit in operation.

IR and UV

As well as visible LEDs, you can also buy LEDs whose light is invisible. This is not as pointless as you might think. Infrared LEDs are used in TV remote controls, and ultraviolet LEDs are

FIGURE 4-6 An RGB LED test schematic

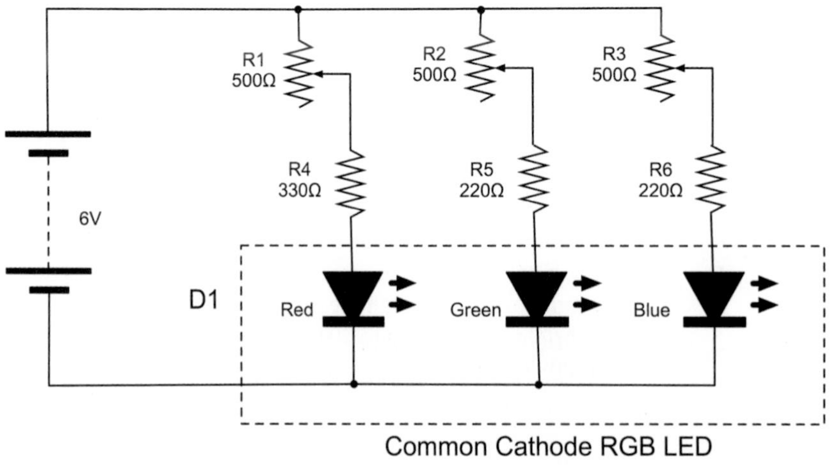

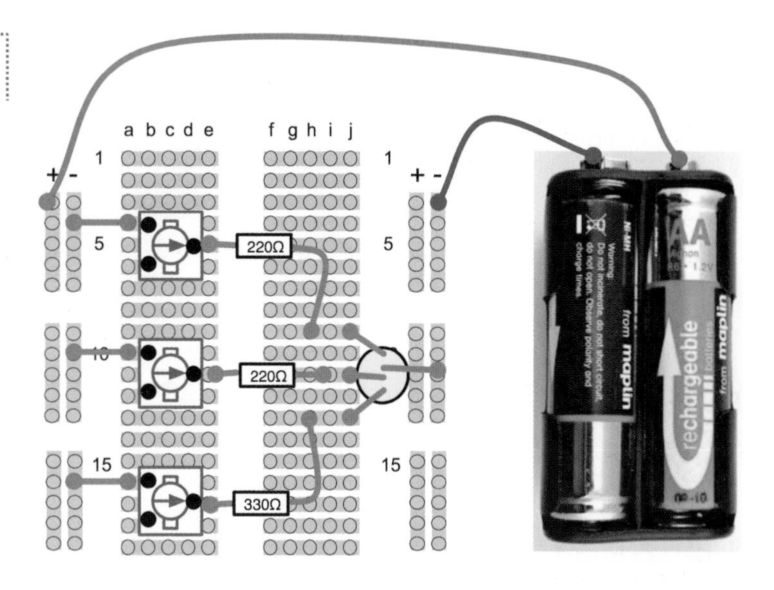

FIGURE 4-7 An RGB LED test breadboard layout

used in specialist applications such as checking the authenticity of bank notes and making people's white clothes light up in clubs.

Use these LEDs just like any other LED. They will have a recommended forward current and voltage and will need a series resistor. Of course, checking that they are working is trickier. Digital cameras are often a little sensitive to infrared and you may see a red glow on the screen.

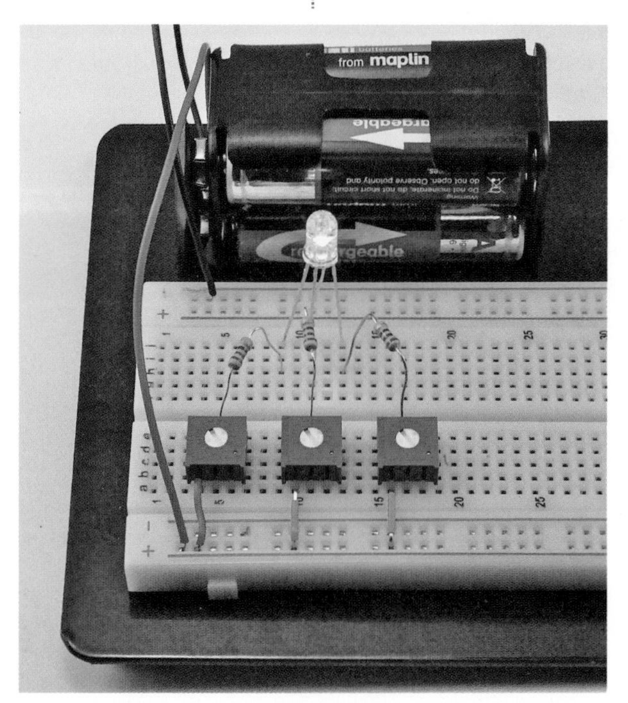

FIGURE 4-8 An RGB test

LEDs for Illumination

LEDs are also finding their way into general household lighting. This has come about because of improvements in LED technology that have produced LEDs with a brightness comparable to incandescent light bulbs—well, they are getting there anyway. Figure 4-9 shows one such high-brightness LED. In this case, it's a 1-W LED, although 3-W and 5-W LED modules are also available.

The cool-looking star shape is thanks to the aluminum heat sink that the LED is attached to. At full power, these LEDs produce enough heat to warrant such a heat sink to disperse the heat into the air.

These LEDs can use a resistor to limit the current, but a quick calculation will show you that you will need quite a high-power resistor. A better approach to using these LEDs is to use a constant current driver, which is the subject of the next section.

FIGURE 4-9 A high-power LED

How to Use a LM317 to Make a Constant Current Driver

Using a resistor to limit the current is all right for small LEDs. However, it is a bit hit or miss since it is very dependent on the LED being used and the power being provided. So for low-power LEDs, where the supply current is not critical it works okay. For high-power LEDs, you can use a series resistor (it will need to be quite high power), but a better way is to use a constant current driver.

As the name suggests, the constant current driver will supply the same current whatever voltage it is supplied with and whatever the forward voltage of the LED. You just set the current and that is how much current will flow through the high-power LED.

A very useful IC that is often used for this purpose is the LM317. This IC is primarily intended as an adjustable voltage regulator, but can easily be adapted for use in regulating current.

This project will start off on breadboard and then we will cut the top off a battery clip and solder the LM317 and resistor to it to make an emergency 1-W LED light.

You Will Need

Quantity	Names	Item	Appendix Code
1		Solderless breadboard	T5
1	D1	1-W white Lumileds LED	S3

Quantity	Names	Item	Appendix Code
3	R1	4.7Ω resistor	K2
1		Battery clip (to destroy)	H2
1		PP3 9V battery	
		Jumper wires	T6

Design

Figure 4-10 shows the schematic diagram for regulating the current to a high-power LED like the one shown in Figure 4-9.

The LM317 is very easy to use in a constant current mode. It will always strive to keep its output voltage at exactly 1.25V above whatever voltage the Adj (adjust) pin is at.

The LED we are going to use is a 1W white light LED. It has an I_f (forward current) of 300mA and a V_f (forward voltage) of 3.4V.

The formula for calculating the right value for R1 for use with the LM317 is:

$$R = 1.25V / I$$

so in this case, $R = 1.25 / 0.3 = 4.2\Omega$

If we used a standard resistor value of 4.7Ω, then this would reduce the current to:

$$I = 1.25V / 4.7\Omega = 266 \text{ mA}$$

Checking the power rating for the resistor, the LM317 will always have 1.25V between Out and Adj. So:

$$P = V \times I = 1.25V \times 266mA = 0.33W$$

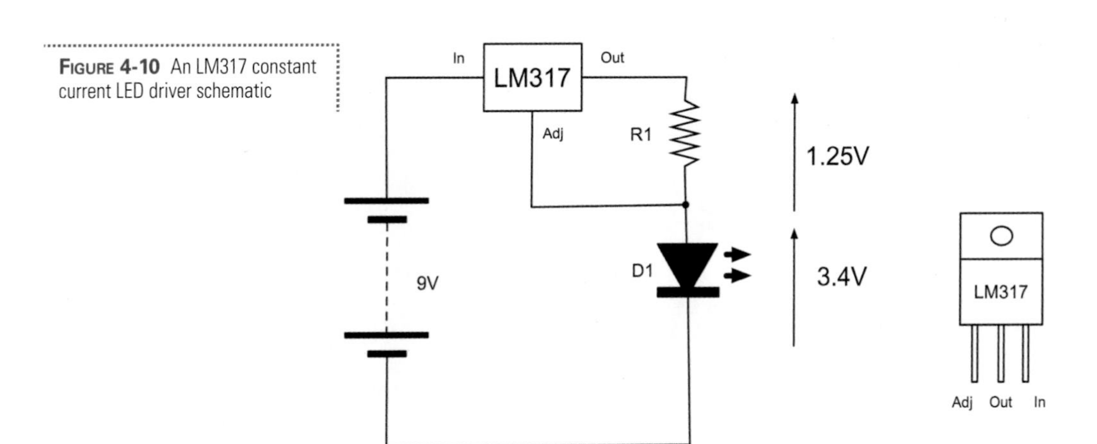

Figure 4-10 An LM317 constant current LED driver schematic

A half-watt resistor will therefore be fine.

The LM317 also needs its input to be about 3V higher than its output to guarantee 1.25V between Adj and the output. This means that a 6V battery would not be quite high enough because the forward voltage is 3.4V. However, we could drive the circuit using a 9V battery or even a 12V power supply without modification, since whatever the input voltage, the current will always be limited to about 260mA.

A quick calculation of the power consumed by the LM317 will reassure us that we are not going to come near exceeding its maximum power rating.

For a 9V battery, the voltage between In and Out will be $9 - (1.25 + 3.4) = 4.35V$. The current is 260mA, so the power is: $4.35 \times 0.26 = 1.13W$.

According to its data sheet, the maximum power handling capability of the LM317 is 20W, and it can cope with a current of up to 2.2A for a supply voltage of less than 15V. So we are fine.

Breadboard

Figure 4-11 shows the breadboard layout for this, and Figure 4-12 displays the actual breadboard. These LEDs are almost painfully bright, so avoid staring at them. When working with them, I cover them with a sheet of paper so I can see when they are on without being temporarily blinded!

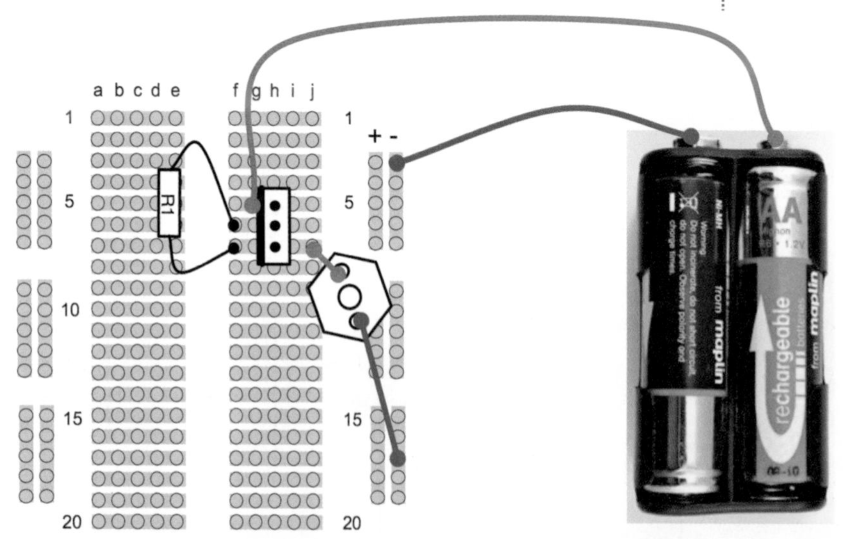

FIGURE 4-11 The LED constant current driver breadboard layout

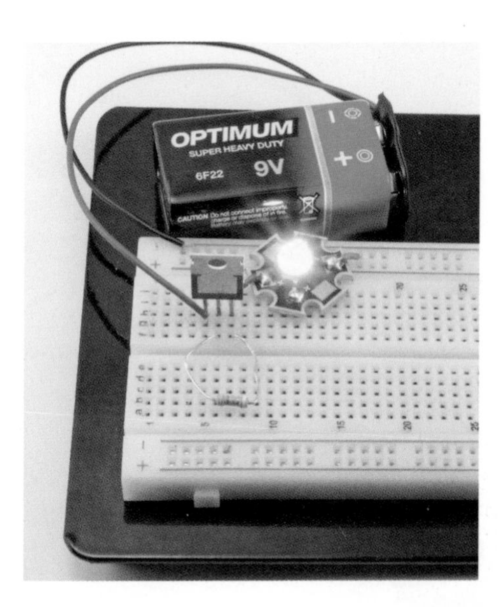

FIGURE 4-12 The LED constant current driver

You will need to solder lengths of solid-core wire to the LED's terminals so it can plug into the breadboard. It is a good idea to leave the insulation on so there is no chance of the bare wires touching the heat sink and shorting.

Construction

We will use this to make a little emergency lantern by hacking a battery clip to build the electronics on top of it so it can be clipped on top of a PP3 battery in the event of a power failure (Figure 4-13).

FIGURE 4-13 Emergency LED lighting

Figures 4-14a thru 4-14d shows the stages involved in soldering this up.

First, remove the plastic from the back of the battery clip using a craft knife. Then, unsolder the exposed leads (Figure 4-14a).

The next step (Figure 4-14b) is to solder the Input lead of the LM317 to the positive terminal of the battery clip. Remember that the positive connector on the clip will be the opposite of the connector on the battery itself, so the positive connector on the clip is the socket-shaped connector. Gently bend the leads of the LM317 apart a little to make this easier.

Now solder the LED in place making sure the cathode of the LED goes to the negative connection on the battery clip (Figure 4-14c).

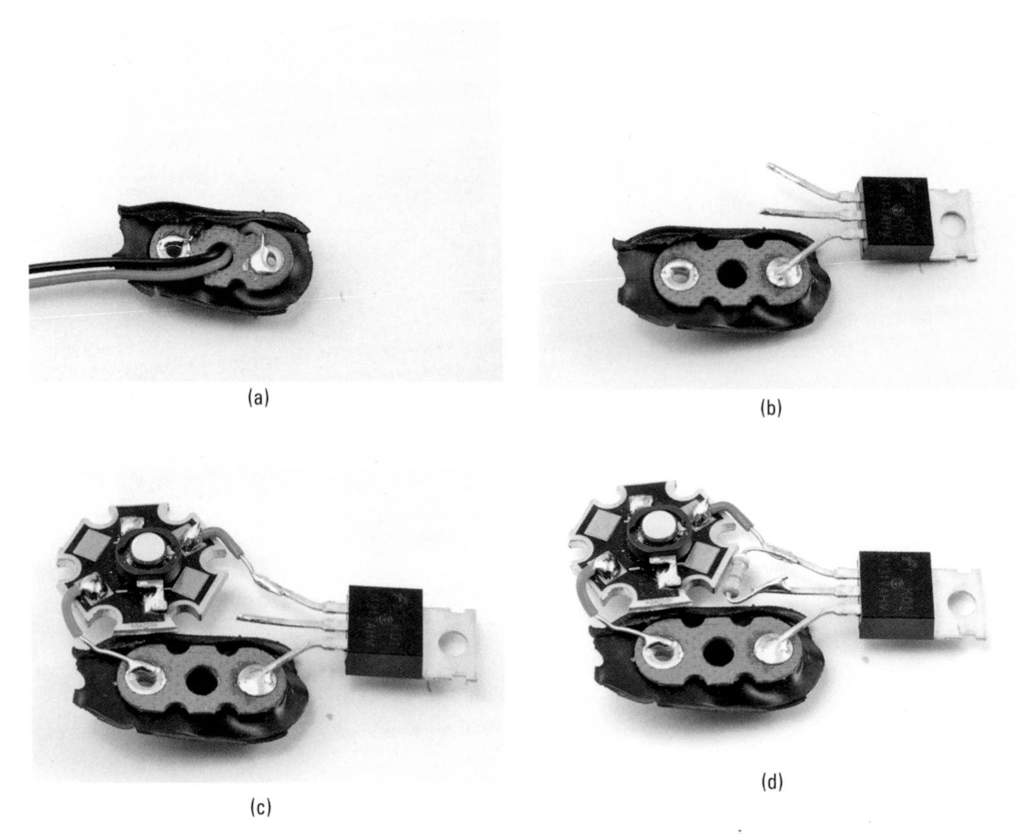
(a) (b) (c) (d)

FIGURE 4-14 Making an emergency 1-W LED light

Finally, solder the resistor across the two topmost leads of the LM317 (Figure 4-14d).

How to Measure the Forward Voltage of an LED

If you want to power lots of LEDs at the same time, it is a good idea to test a few of the LEDs you actually intend to use and measure the forward voltage at the current you intend to use. Figure 4-15 shows how you would do this.

Figure 4-15a shows the schematic diagram. A variable resistor is used to vary the current through the LED and when the desired forward current is set, the voltage can be read.

The current and voltage do not have to be read at the same time, but if you do have two meters, this does make life easier.

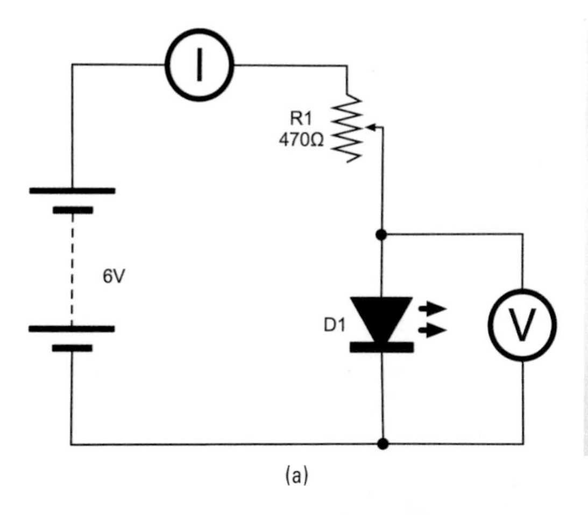

(a)

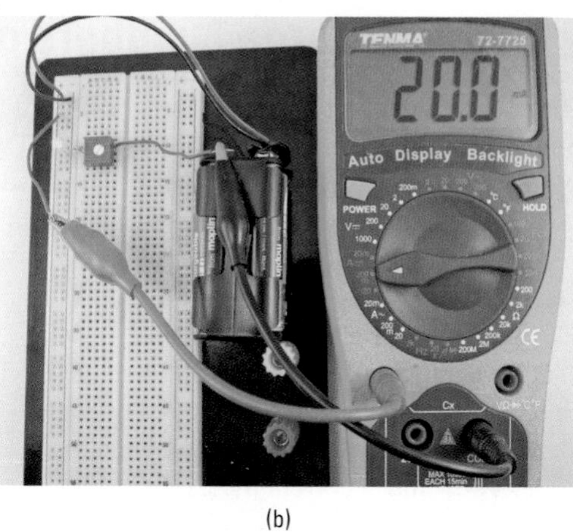

(b)

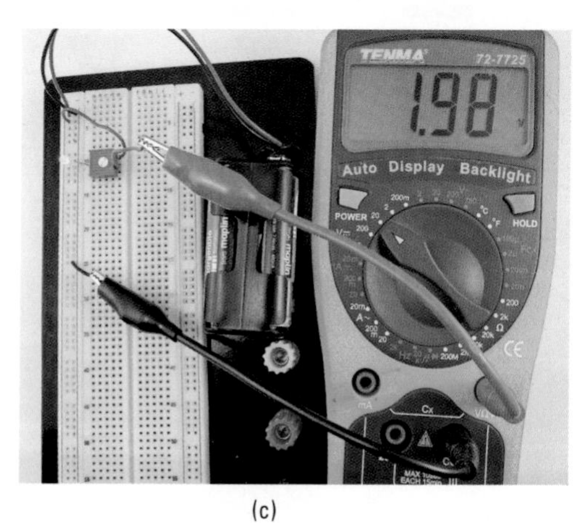

(c)

FIGURE 4-15 Measuring the forward voltage of an LED

Set the variable resistor to its middle point and wire up the breadboard as shown in Figure 4-15b. You will probably have to move the positive lead of your multimeter to a different socket when measuring current. Select a current range of 200mA DC. Now adjust the variable resistor until the current reads 20mA.

We can now measure the voltage across the LED. To do this, first disconnect the multimeter, put the positive multimeter back in the right socket for measuring voltage, and then change the range to 20V DC. Wire it up as shown in Figure 4-15c and measure the voltage. In this case, it was 1.98V.

You Will Need

Quantity	Names	Item	Appendix Code
1		Solderless breadboard	T5
1	D1	LED	K1
3	R1	500Ω trimpot	R3
		Jumper wires	T6
1		4 × AA battery holder	H1
1		Battery clip	H2
4		AA batteries	

How to Power Large Numbers of LEDs

If you use something like a 12V power supply, you can put a number of LEDs in series, with just one LED. In fact, if you know the forward voltages fairly accurately, and the power supply is well regulated, you can get away without any series resistor at all.

So, if you have fairly standard LEDs that have a forward voltage of 2V, then you could just put six of them in series. However, it will not be terribly easy to predict how much current the LEDs will take.

A safer approach is to arrange the LEDs in parallel strings, each string having its own current-limiting resistor (Figure 4-16).

Although the math for this isn't too hard, there can be a fair bit of it, so you can save yourself a lot of time by using an online calculator like http://led.linear1.org/led.wiz (Figure 4-17).

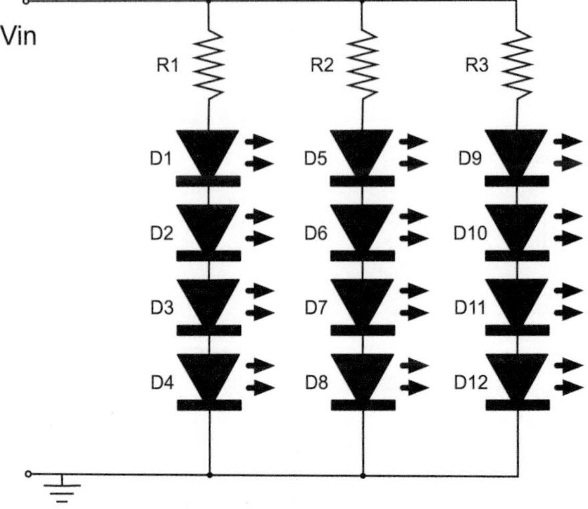

FIGURE 4-16 Powering multiple LEDs

LED series/parallel array wizard

The LED series/parallel array wizard is a calculator that will help you design large arrays of LEDs. The LED calculator was great for single LEDs--but when you have several, the wizard will help you arrange them in a series or combined series/parallel configuration. The wizard determines the current limiting resistor value for each portion of the array and calculates power consumed. All you need to know are the specs of your LEDs and how many you'd like to use.

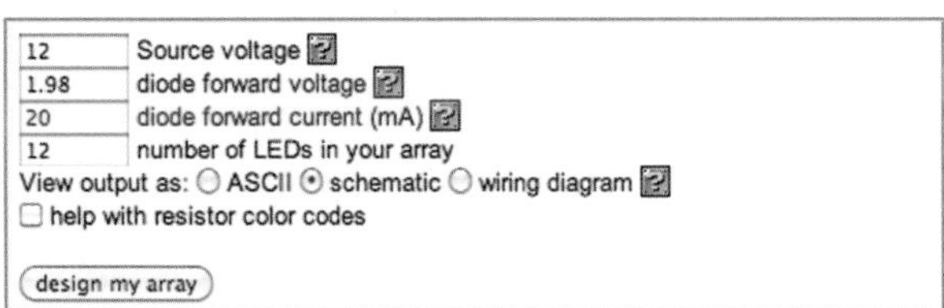

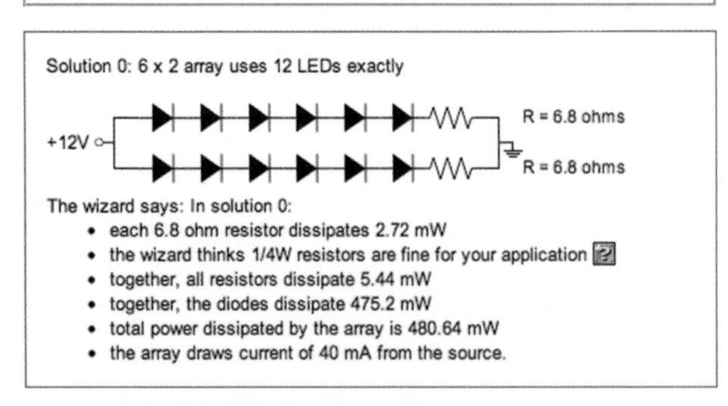

FIGURE 4-17 The LED wizard

In this particular designer, you enter the source voltage for the overall supply, the LED forward voltage, the desired current for each LED, and the number of LEDs you want to light. The wizard then does the math and works out a few different layouts.

One consideration is that where you have a string of LEDs in series, if any of the LEDs fail, then all the LEDs will be off.

How to Make LEDs Flash

The 555 timer IC is a useful little IC that can be used for many different purposes, but is particularly convenient for making LEDs flash or generating higher frequency oscillations suitable for making audible tones (see Chapter 9).

We are going to make this design on breadboard and then transfer it to a more permanent home on a bit of stripboard.

You Will Need

Quantity	Names	Item	Appendix Code
1		Solderless breadboard	T5
1	D1	LED red	K1
1	D1	LED green	K1
1	R1	1kΩ resistor	K2
1	R2	470kΩ resistor	K2
2	R3, R4	220Ω resistor	K2
1	C1	1μF capacitor	K2
1	IC1	555 timer	K2
		Jumper wires	T6
1		4 × AA battery holder	H1
1		Battery clip	H2
4		AA batteries	

Breadboard

The schematic for the LED flasher is shown in Figure 4-18.

The breadboard layout is shown in Figure 4-19. Make sure you have the IC the right way up. There will be a notch in the IC body next to the top (pins 1 and 8). The capacitor and LEDs must both be the correct way around, too.

Figure 4-20 shows the finished breadboard. You should find that the LEDs alternate, each staying on for about a second.

Now that we know that the design is right and everything works, try swapping out R2 with a 100kΩ resistor and notice the effect on the flashing.

The 555 timer is a very versatile device, and in this configuration it oscillates at a frequency determined by the formula:

frequency = 1.44 / ([R1 + 2 * R2] * C)

where the units of R1, R2, and C1 are in Ω and F. Plugging in the values for this design, we get:

frequency = 1.44 / ([1000 + 2 * 470000] * 0.000001) = 1.53 Hz

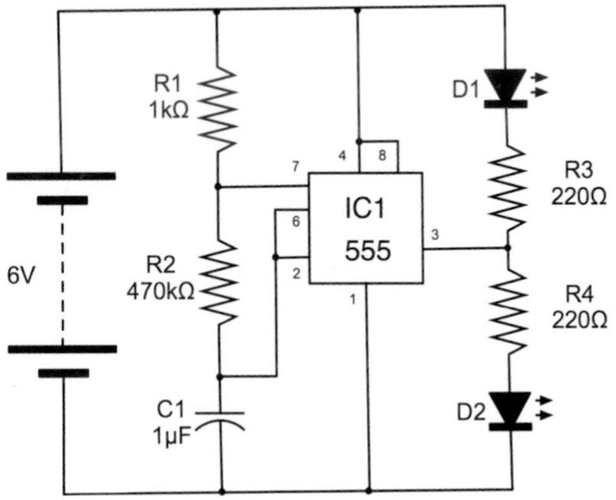

Figure 4-18 The LED flasher schematic

One hertz (or Hz) means one oscillation per second. When we use the 555 timer in a later chapter to generate an audible tone, we will be using the same circuit to generate a frequency in the hundreds of hertz.

As with so many electronic calculations, there are also online calculators for the 555 timer.

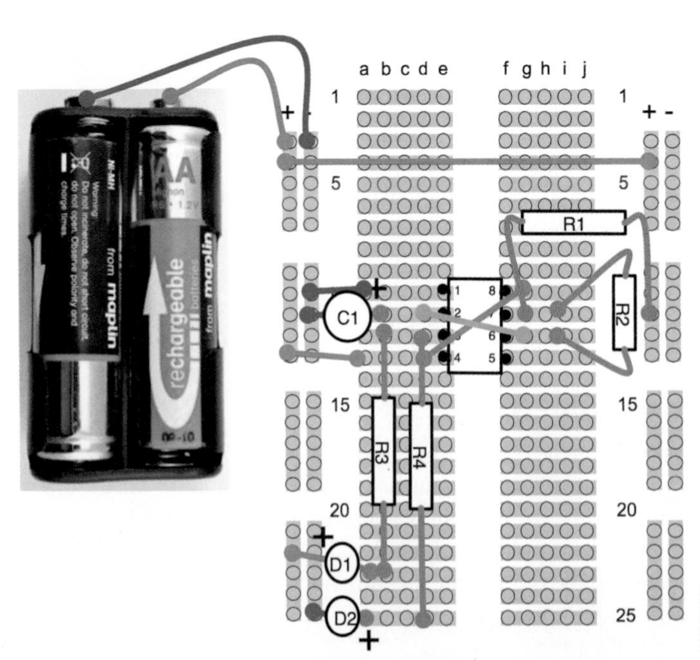

Figure 4-19 The LED flasher breadboard layout

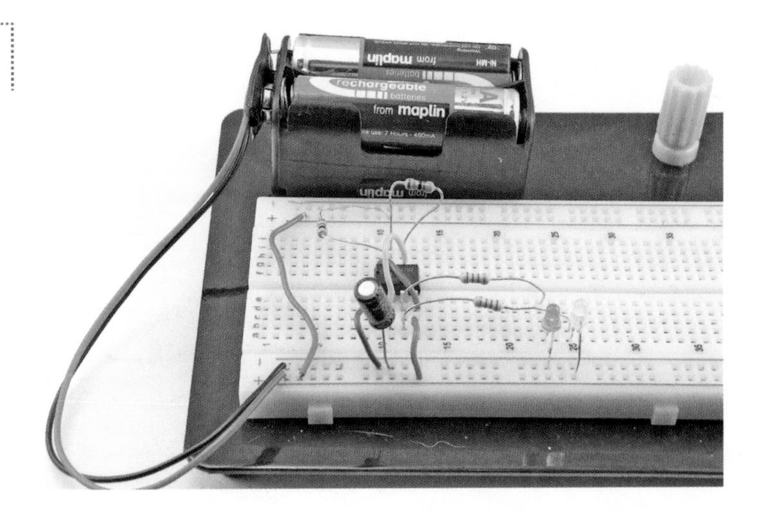

FIGURE 4-20 The LED flasher on breadboard

How to Use Stripboard (LED Flasher)

Breadboard is very useful for trying things out, but not so useful as a permanent home for your electronics. The problem is that the wires tend to fall out, and it's all a bit big and bulky.

Stripboard (Figure 4-21) is a bit like general-purpose printed circuit board. It is a perforated board with conductive strips running underneath, rather like breadboard. The board can be cut to the size you need and components and wires soldered onto it.

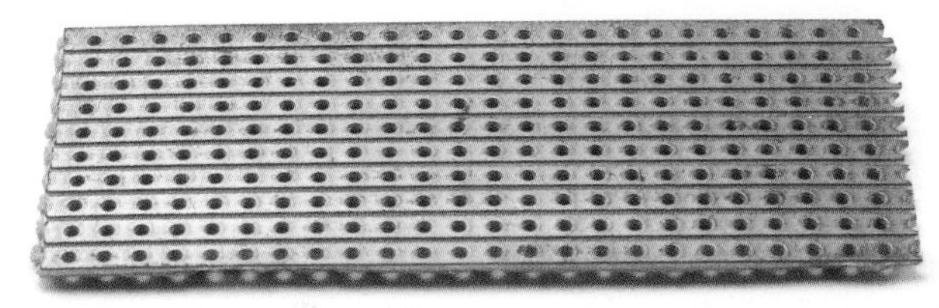

FIGURE 4-21 Stripboard

Designing the Stripboard Layout

Figure 4-22 shows the final stripboard layout for the LED flasher that we made in the previous section. It is not easy to explain how we got to this from the schematic and breadboard layout. There is a certain amount of trial and error, but there are a few principals you can follow to try and make it easier.

The first is to use a drawing tool with a stripboard template. For Mac users, with OmniGraffle, a template is available for download from the book's web site (www.hackingelectronics .com). There is also an image file that can be printed out and used as a template to sketch out the design.

The Xs underneath the IC are breaks in the track, which we will make with a drill bit. One of the goals of a good stripboard

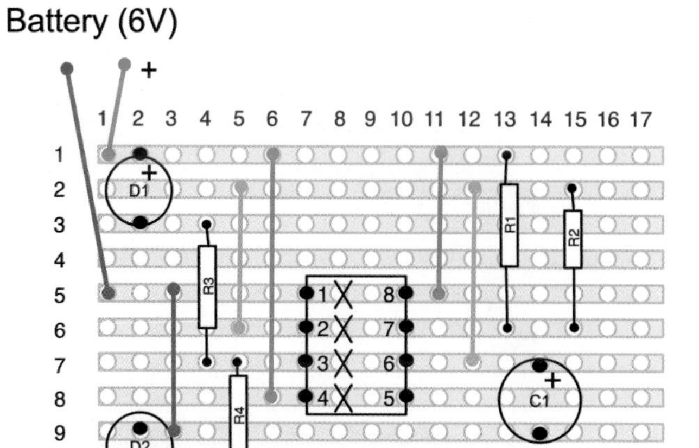

FIGURE 3-22 An LED flasher stripboard layout

layout is to try and avoid making too many breaks in the track. Breaks are unavoidable for an IC like this. If we did not make them, pin 1 would be connected to pin 8, pin 2 to pin 7, and so on, and nothing would work.

The colored lines on the board are linking wires. So, for instance, from the schematic diagram of Figure 4-18, we can see that pins 4 and 8 of the IC should be connected together and both go to the positive supply. This is accomplished by the two red linking wires. Similarly, pins 2 and 6 need to be connected together. This is accomplished by using the orange leads.

Although logically the stripboard layout is the same as the schematic, the components are in rather different places. The LEDs are on the left in the stripboard layout and on the right on the schematic. It is not always like this, and it's easier if they are similar, but in this case the left-hand pins of the IC include the output pin 3 that the LEDs need, and the pins connected to R1, R3, and C1 are all on the right-hand side of the IC.

Try making a stripboard layout from the schematic, you may well come up with a different and better layout than the one I produced.

The steps I went through in designing this layout are as follows:

1. Place the IC fairly centrally, with a bit more room above than below and with pin 1 uppermost (convention).
2. Find a good place for R3 and R4 to be put so the strips are at least three holes apart for one resistor lead, when the other lead of each resistor goes to pin 3.

3. Pick the top track of the stripboard to be +V so it can be close to the positive end of one of the LEDs
4. Pick row 5 to be the ground connection. This way it can run straight on to pin 1 of the IC.
5. Add a link wire from row 5 to row 9 to provide the negative connection for the LED D2.
6. Put a jumper wire from pin 4 of the IC to row 1 (+V).

Turning now to the right-hand side of the board:

1. Put a jumper in connecting pin 8 of the IC to row 1 (+V).
2. R1 and R2 both have one end connected to pin 7, so put them side by side with the far end of R1 going to row 1 (+V).
3. R2 needs to connect to pin 6, but pin 6 and pin 7 of the IC are too close together for the resistor to lie flat, so take that lead up to the unused row 2 instead, then put jumpers from row 2 down to both pin 6 and pin 2 of the IC.
4. Finally, C1 needs to go between pin 6 (or pin 2, but 6 is easier) and GND (row 9).

A good way of checking that you have made all the connections you need is to print off the schematic and then go through each connection on the stripboard and check off its counterpart on the schematic.

This may all seem a little like magic, but try it. It's not as hard to do as it is to describe.

You Will Need

You will need all the components listed in the section "How to Make LEDs Flash," plus the following items.

Quantity	Item	Appendix Code
1	Stripboard 10 strips by 17 holes	H3
1	Soldering kit	T1
1	Drill bit (1/8 inch)	

Before we start soldering, it is worth considering what kind of LEDs you want to use for this project. You may decide to use higher-brightness LEDs or power the project from a lower voltage. If you do decide on this, recalculate the values for R3 and R4 and try it out on the breadboard layout. The 555 timer IC needs a supply voltage between 4.5V and 16V, and the output can supply up to 200mA.

Construction

Step 1. Cut the Stripboard to Size

There is no point in having a large bit of stripboard with just a few components on it, so the first thing we need to do is cut the stripboard to the right size. In this case, it's ten strips each of 17 holes. Stripboard doesn't actually cut very well. You can use a rotary tool, but wear a mask because the dust from the stripboard is nasty and you really do not want it in your lungs. I find the easiest way to cut stripboard is actually to score it with a craft knife and metal ruler on both sides and then break it over the edge of your work surface.

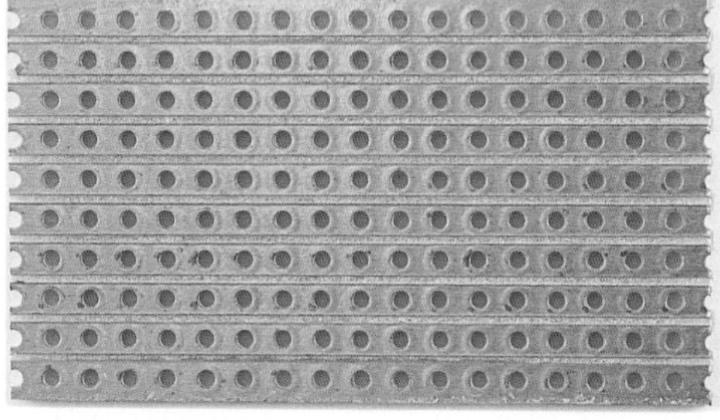

FIGURE 4-23 A stripboard cut to size

Score it across the holes, not between them. When the board is cut, the copper underside will look like Figure 4-23.

Step 2. Make the Breaks in the Tracks

A good tip is to use a permanent marker and put a dot in the top left corner. Otherwise, it is very easy to get the board turned around, resulting in breaks and links being put in the wrong place.

To make the breaks, count the position in rows and columns of the break from the top of the board layout and then push a bit of wire so you can identify the right hole on the copper side of the stripboard (Figure 4-24a). Using a drill bit, just "twizzle" it between thumb and forefinger to cut through the copper track. It usually only takes a couple of twists (Figure 4-24b and c).

When you have cut all four breaks, the bottom side of the breadboard should look like Figure 4-25. Check very carefully that none of the burrs from the copper have lodged between the tracks and that the breaks are complete. Photographing the board and then zooming in is a great way of actually checking the board.

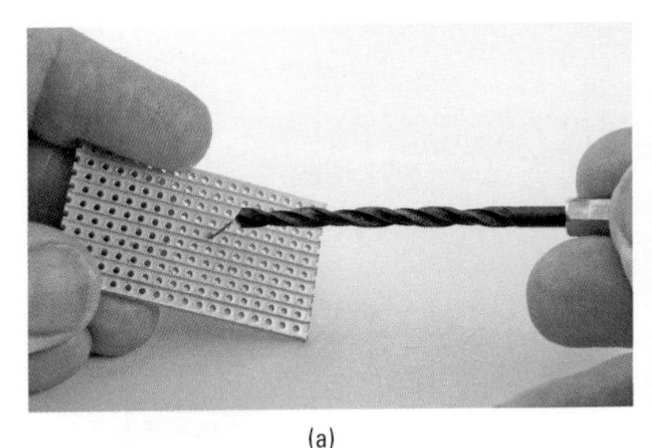
(a)

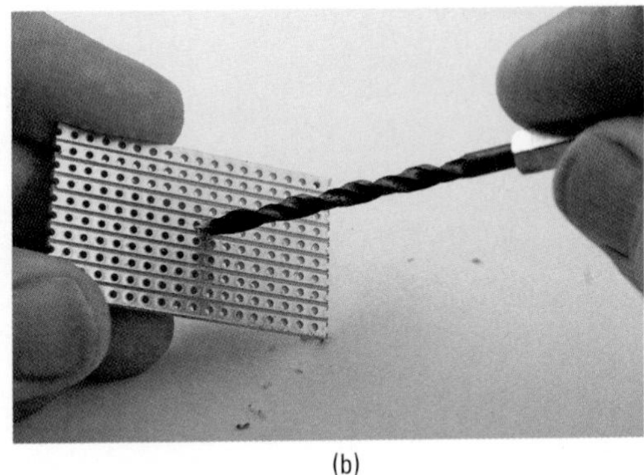
(b)

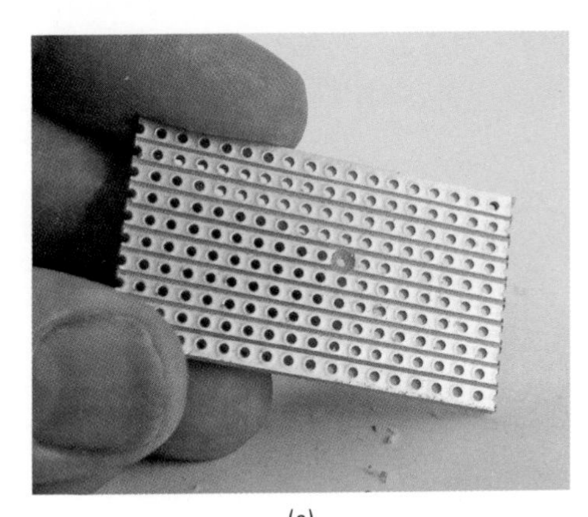
(c)

FIGURE 4-24 Cutting a track on stripboard

Step 3. Make the Wire Links

A golden rule of any type of circuit board construction, including when using stripboard, is to start with the lowest-lying components. This is so that when you turn the board on its back to solder it, the thing being soldered will stay in place through the weight of the board.

In this case, the first thing to solder is the links.

Strip and cut solid-core wire to slightly longer than the length of the link. Bend it into a U-shape and push it through the holes at the top, counting the rows and columns to get the right holes (Figure 4-26a). Some people get very skilled at bending the wires with pliers to just the right length. I find it easier to bend the wires with a bit of a curve so they will kind of squash into the right holes. I find this easier than trying to get the length just right from the start.

Turn the board over (see how the wire is held in place) and solder the wire by holding the iron to the point where the wire emerges from the hole. Heat it for a second or two and then apply the solder until it slows along the track, covering the hole and wire (Figure 4-26b and c).

Repeat this process for the other end of the lead and then snip off the excess wire (Figure 4-26d and e).

When you have soldered all the links, your board should look like the one in Figure 4-27.

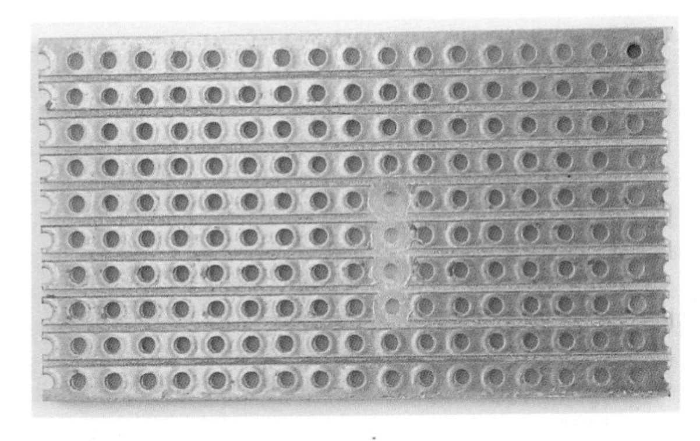

FIGURE 4-25 The stripboard with breaks cut

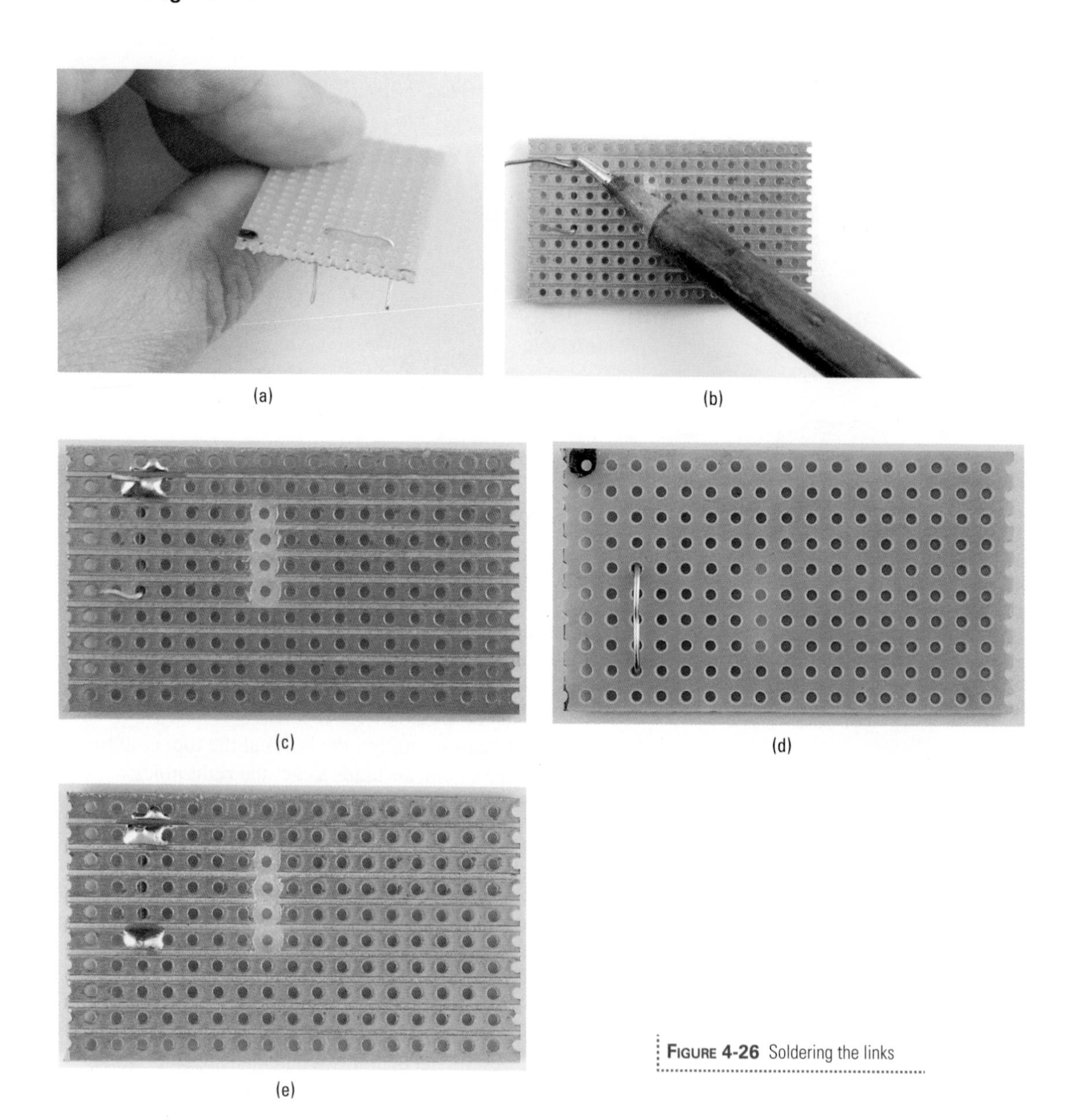

(a) (b) (c) (d) (e)

FIGURE 4-26 Soldering the links

Step 4. Resistors

The resistors are the next lowest components to the board, so solder these next, in the same way as you did the links. When they are all soldered, the stripboard should look like Figure 4-28.

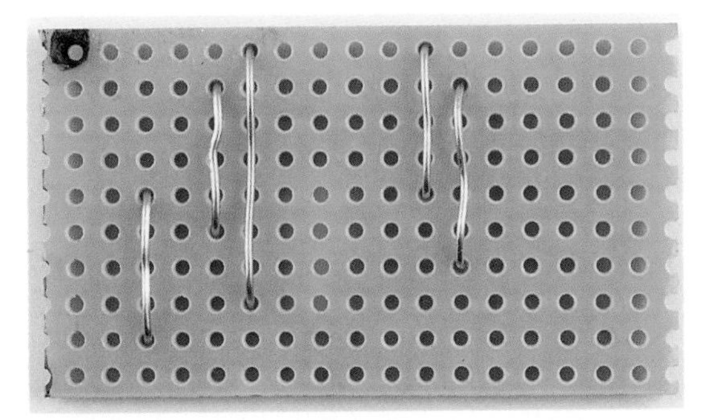

Figure 4-27 The stripboard with all its links

Step 5. Solder the Remaining Components

Next, solder the LED, capacitor (which can be laid on its side as shown in Figure 4-29), and finally the LEDs and connectors to the battery clip.

That's it. Now it's time for the moment of truth. Before you plug it in, do a very careful inspection for any shorts on the underside of the board.

If everything seems in order, connect the battery clip to the battery.

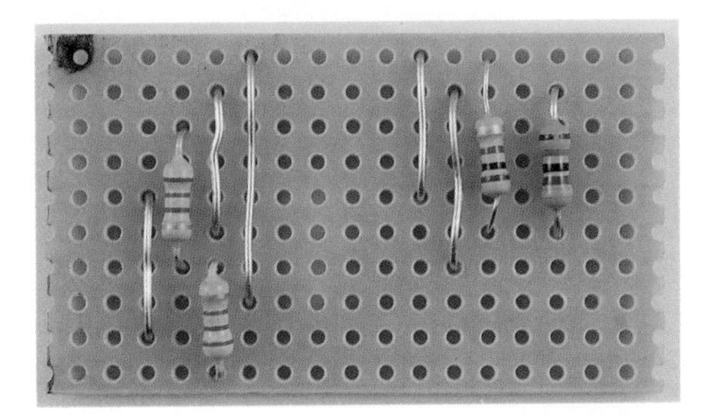

Figure 4-28 The stripboard with resistors in place

Troubleshooting

If it does not work, immediately unplug it and go back though and check everything, especially that the LEDs, IC, and capacitor are the correct way around. Also check that the batteries are okay.

Figure 4-29 The LED flasher on the stripboard

How to Use a Laser Diode Module

Lasers are best bought as laser modules. The difference between a laser module and a laser diode is that the module includes a laser diode as well as a lens to focus the beam of laser light and a drive circuit to control the current to the laser diode.

If you buy a laser diode, you will have to do all this yourself.

A laser diode module, such as the 1-mW module shown in Figure 4-30, comes with a datasheet that says it needs to be supplied with 3V. So, all you need to do is find it a 3V battery and connect it up.

FIGURE 4-30 A laser diode module

Hacking a Slot Car Racer

Slot cars are a lot of fun, but could be improved by adding headlights and working brake lights to the car (Figure 4-31).

LEDs are just the right size to be fitted front and back into a slot car.

FIGURE 4-31 A modified slot car racer

You Will Need

You will need the following items to add lights to your slot car.

Quantity	Name	Item	Appendix Code
1		Slot car racer for modification	
1	D1	1N4001	S5, K1
2	D2, D3	High-brightness white LED 5rmm	S2
2	D4, D5	Red LED 5mm	S11
4	R1–4	1kΩ resistor	K2
1	C1	1000μF 16V-capacitor	C1
		Red, yellow, and black hookup wire	T7, T8, T9
1		* Two-way header plug and socket	

* I used a scavenged two-way header socket and plug to make it easier to work on the two halves of the car. This is not essential.

The slot car used here was part of a build-your-own slot car that has plenty of room inside for the electronics. Plan ahead and make sure you can fit everything in.

Storing Charge in a Capacitor

To make the brake lights stay on for a few moments after the car has stopped, you will need a capacitor to store charge.

If you think back to the idea of electricity as water flowing in a river, then a capacitor is a bit like a storage tank. Figure 4-32 shows how a capacitor can be used to store charge.

Figure 4-32a shows a tank (c1) being filled with water through A. Throughout this, water will also flow along the top and drive a water wheel, turning electrical energy into motion, a bit like how a light bulb or LED turns electrical energy into light. The water falls out of the bottom, returning to ground. Imagine a pump (like a battery) pulling the water back up for another circuit. If the water stops flowing into C1 through A, then C1 will still be full of water that will keep the water wheel turning until the water level in C1 drops below that of the outlet of the water well.

Figure 4-32b shows the electronic equivalent of this circuit. While the voltage at A is higher than GND, C1 will fill with charge and the light will be lit.

When the voltage at A is disconnected, the capacitor will discharge through the light bulb, lighting it. As the level of voltage drops in the capacitor, the bulb will gradually dim until it goes out as it reaches GND.

FIGURE 4-32 A capacitor as a tank

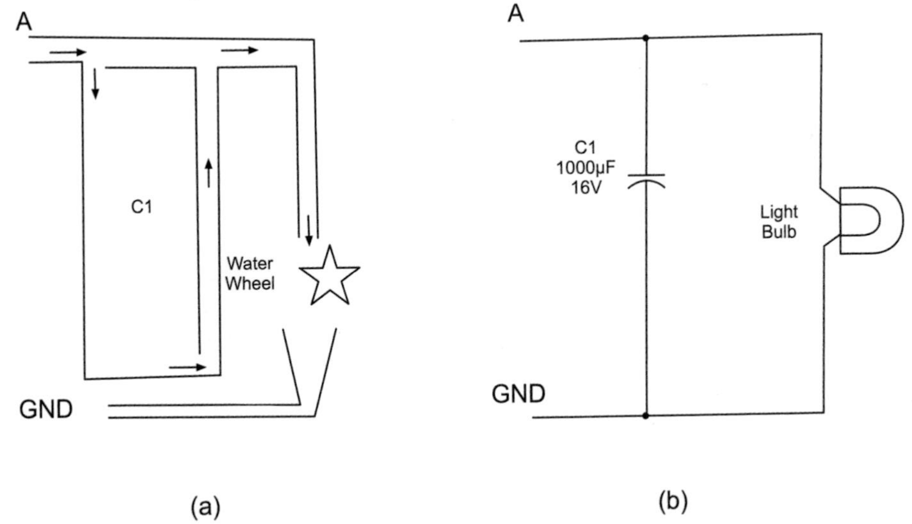

On the face of it, you can think of capacitors as being a bit like batteries. Both store charge. However, there are some very important differences.

- Capacitors only store a tiny fraction of the charge that a battery of the same size can store.
- Batteries use a chemical reaction to store electrical energy. This means their voltage remains relatively constant until they are spent, at which time it falls off rapidly. Capacitors, however, drop evenly in voltage as they discharge, just like the level of water decreasing in a tank.

Design

Figure 4-33 shows the schematic diagram for this modification.

The headlights (D2 and D3) are powered all the time from the slot car's connection to the track, so whenever the motor is running the LEDs will light.

The brake lights are more interesting. These will automatically come on when the car stops, and then go off after a few seconds. To do this, we make use of a capacitor C1.

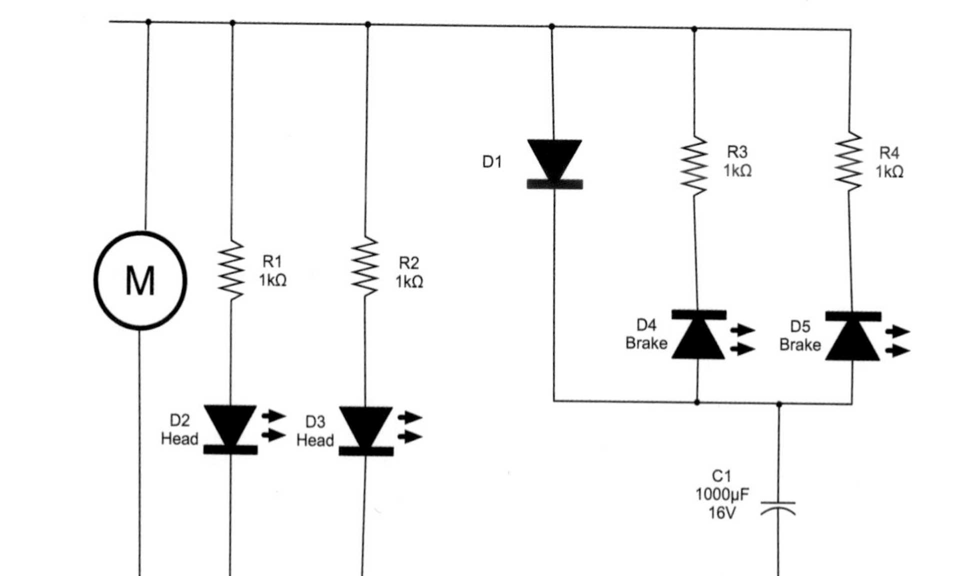

FIGURE 4-33 A schematic diagram for the slot car modification

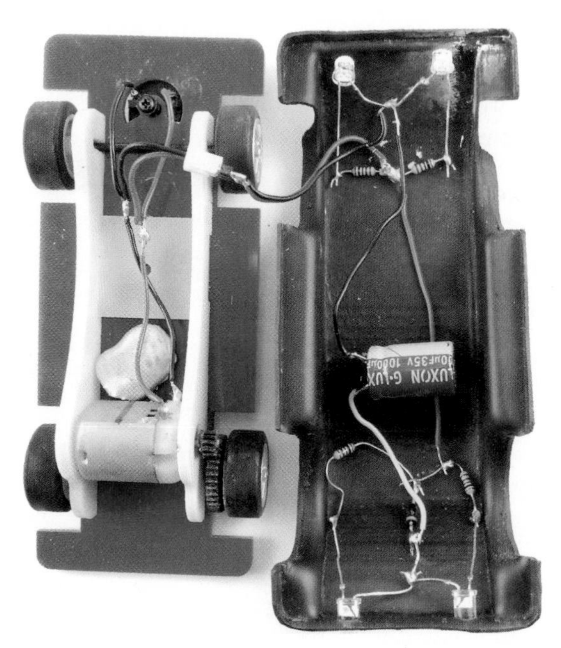

Figure 4-34 The components inside the car

When the car is powered, C1 will be charged through D1. At this point, the brake lights D4 and D5 will not be lit because they will be reverse-biased—that is, the voltage going in from the car tracks will be higher than the voltage at the top of the capacitor.

When you release the trigger on the controller, there will be no voltage coming in. Now the voltage at the top of the capacitor will be higher than the voltage coming in, so the capacitor will discharge through D4 and D5, making them light.

Construction

Figure 4-34 shows how the components are laid out in the two halves of the car.

How you lay these out in your vehicle may vary depending on how much space you have.

Holes were drilled in the case to take the 5mm LEDs. The LEDs are a snug fit in the holes and stay in place without any glue.

Figure 4-35 shows a wiring diagram for the arrangement that makes it easier to see what is going on.

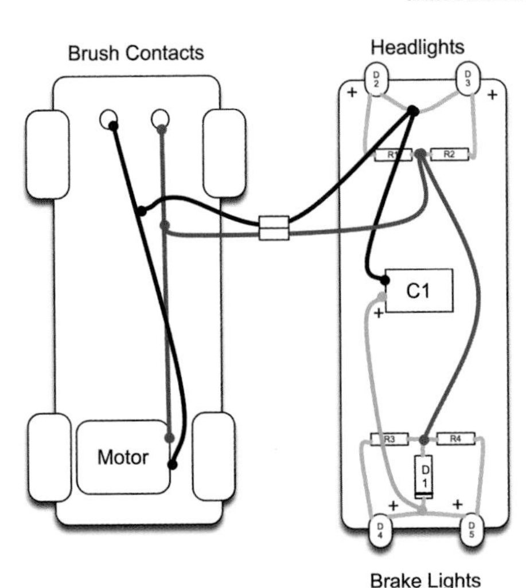

Figure 4-35 The wiring diagram for the modified car

Use your multimeter of the 20V range to identify which is the positive power connector on the contacts at the front of the car. This contact is connected to the red lead.

The longer leads of the LEDs are the positive connections, and the capacitor's negative lead should be marked with a "-".

The optional connector makes it easier to work on the two halves of the car separately.

Testing

Testing really just involves trying out the car on the track. If the headlight LEDs are not on as soon as you squeeze the trigger on the controller, check the wiring, paying special attention to the polarity of the LEDs.

Summary

We have learned how to use LEDs in this chapter, as well as picked up some good building skills so we can make our creations a bit more permanent using stripboard.

In the next chapter, we will examine sources of power, including batteries, power supplies, and solar panels. We will also look at how to select the right kind of battery, repurpose old rechargeable batteries, and use them in our projects.

5

Batteries and Power

Everything that you make or adapt is going to need to get its power from somewhere. This might be from a household electricity adapter, solar panels, rechargeable batteries of various sorts, or just standard AA batteries.

In this chapter, you will find out all about batteries and power, starting with batteries.

I use the word "battery" to describe both batteries and cells. Strictly speaking, a battery is a collection of cells wired one after the other to give the desired voltage.

Selecting the Right Battery

There are many types of battery on the market. So, to simplify things, in this chapter we will just look at the most common types of battery, those that are readily available and that will be used in most of the devices in this book.

Battery Capacity

Whether single-use or rechargeable, batteries have a capacity—that is, they hold a certain amount of electricity. Manufacturers of single-use batteries often don't specify this capacity in the batteries you buy from a supermarket. They just label them heavy duty / light duty, and so on. This is a little like having to buy milk and being given the choice of "big bottle" or "small bottle" without being able to see how big the bottle is or be told how many pints or liters it contains. One can speculate as to the reasons for this. One reason might be that battery producers think the public isn't intelligent enough to understand a stated battery capacity. Another might be that the longer a battery is on the shelf, the more its capacity shrinks. Still another is that the capacity actually varies a lot with the current drawn from the battery.

Anyway, if a battery manufacturer is kind enough to tell you what you are buying, the capacity figure will be stated in Ah or mAh. So a battery that claims to have a capacity of 3000mAh (typical of a single-use alkaline AA cell) can supply 3000mA for one hour. Or, alternatively, 3A for an hour. But it doesn't have to draw 3A. If your project only uses 30mA, you can expect the battery to last 100 hours (3000/30). In truth, the relationship is not quite that simple, because as you draw more current, the capacity decreases. Nevertheless, this will do as a rule of thumb.

Maximum Discharge Rate

You cannot take a tiny battery like a CR2032 with a capacity of 200mAh and expect to power a big electric motor at 20A for 1/100 of an hour (six minutes). There are two reasons for this. First, all batteries actually have an internal resistance. So, it is as if there is a resistor connected to one of the terminals. This varies depending on the current being drawn from the battery, but may be as high as a few tens of ohms. This will naturally limit the current.

Second, when a battery is discharged too quickly, by too high a current, it gets hot—sometimes very hot, sometimes "on fire" hot. This will damage the battery.

Batteries therefore also have a safe discharge rate, which is the maximum current you can safely draw from it.

Single-Use Batteries

Although somewhat wasteful, sometimes it makes sense to use single-use batteries that cannot be recharged. You should consider single-use batteries if:

- The project uses very little power, so they will last a long time anyway.
- The project will never be close to someplace where it can be charged up.

Table 5-1 shows some common single-use batteries. These figures are typical values and will vary a lot between actual devices.

Especially when it comes to the maximum discharge rate, you may get away with a lot more, or the battery may fail or get very hot with considerably less. It will also depend on how well ventilated a box they are in, as heating under high currents is a big problem.

So in the spirit of hacking electronics, spend less time planning and more time trying. See how hot it gets and how long it lasts. After all, we are having fun here, not designing a product.

Type		Typical Capacity	Voltage	Max. Discharge Current	Features	Common Uses
Lithium button cell (e.g., CR2032)		200mAh	3V	4mA with pulses up to 12mA	High temperature range (–30 to 80°C); small	Low-power devices; RF remote controls; LED key ring lights, etc.
Alkaline PP3 battery		500mAh	9V	800mA	Low cost; readily available	Small portable electronic devices; smoke alarms; guitar pedals
Lithium PP3		1200mAh	9V	400mA pulses up to 800mA	Expensive; light; high-capacity	Radio receivers
AAA cell		800mAh	1.5V	1.5A continuous	Low-cost; readily available	Small motorized toys; remote controls
AA cell		3000mAh	1.5V	2A continuous	Low cost; readily available	Motorized toys
C cell		6000mAh	1.5V	Probably get away with 4A	High-capacity	Motorized toys; high-powered flashlights
D cell		15,000mAh	1.5V	Probably get away with 6A	High-capacity	Motorized toys; high-powered flashlights

TABLE 5-1 Single-Use Battery Types

Note Some of the photographs in the table are of branded batteries. The figures shown are for batteries of that type, not specifically the batteries listed.

Roll Your Own Battery

A single-cell battery with a voltage of just 1.5V is probably not going to be of any use. You will normally need to put a number of these cells in series (end to end) to produce a battery of the desired voltage.

When you do this, you do not increase the capacity. If each cell was 2000mAh, then if you put four 1.5V batteries in series, the capacity would still be 2000mAh, but at 6V rather than 1.5V.

Battery holders such as the one shown in Figure 5-1 are a great way of doing this. Look closely at how the battery holder is constructed and you'll notice how the positive of one battery is connected to the negative of the next, and so on.

This holder is designed to take six AA batteries so as to produce an overall voltage of 9V. Battery holders like this are available to take two, four, six, eight, or ten cells, both in AA and AAA.

FIGURE 5-1 A battery holder

Another advantage of using a battery holder is that you can use rechargeable batteries instead of single-use batteries. However, rechargeable cells normally have a lower voltage, so you have to take this into account when calculating the overall voltage of your battery pack.

Selecting a Battery

Table 5-2 should help you decide on a suitable battery for your project. There is not always a best answer to the question "Which battery should I use?" and this table is definitely in the territory of rules of thumb.

You should also do the math and include how frequently the battery will need replacing.

	Voltage			
Power	**3V**	**6V**	**9V**	**12V**
Less than 4mA (short bursts) or 12mA continuous	Lithium button cell (e.g., CR2032)	2 × Lithium button cell (e.g., CR2032)	PP3	Unlikely
Less than 3A (short bursts) or 1.5A continuous	2 × AAA battery pack	4 × AAA battery pack	6 × AAA battery pack	8 × AAA battery pack
Less than 5A (short bursts) or 2A continuous	2 × AAA battery pack	4 × AAA battery pack	6 × AAA battery pack	8 × AAA battery pack
Even more	2 × C or D battery pack	4 × C or D battery pack	6 × C or D battery pack	8 × C or D battery pack

TABLE 5-2 Selecting a Single-Use Battery

Rechargeable Batteries

Rechargeable batteries can provide both cost and green benefits over single-use batteries. They are available in different types and in different capacities. Some, such as rechargeable AA or

AAA batteries, are designed as replacements for single-use batteries, and you remove them to charge in a separate charger. Other batteries are intended to be built into your project so all you have to do is plug a power adapter into your project to charge the batteries without removing them. The advent of cheap, high-capacity, low-weight lithium polymer (LiPo) batteries has made this a common approach for many items of consumer electronics.

Table 5-3 shows some commonly used types of rechargeable batteries.

Although there are many more types than this, these are the most commonly used batteries. Each type of battery has its own needs when it comes to charging, and we will look at each in later sections.

Type	Typical Capacity	Voltage	Features	Common Uses
NiMH button cell pack	80mAh	2.4 or 3.6V	Small	Battery backup
NiMH AAA cell	750mAh	1.25V	Low cost	Replacement for single-use AAA cell
NiMH AA cell	2000mAh	1.25V	Low cost	Replacement for single-use AA cell
NiMH C cell	4000mAh	1.25V	High capacity	Replacement for single-use C cell
Small LiPo cell	50mAh	3.7V	Low cost; high capacity for weight and size	Micro-helicopters
LC18650 LiPo cell	2200mAh	3.7V	Low cost; high capacity for weight and size; slightly bigger than AA	High-power flashlights; Tesla Roadster (yes, really—about 6800 of them)
LiPo pack	900mAh	7.4V	Low cost; high capacity for weight and size	Cell phones, iPods, etc.
Sealed lead–acid battery	1200mAh	6/12V	Easy to charge and use; heavy	Intruder alarms; small electric vehicles / wheelchairs

TABLE 5-3 Rechargeable Batteries

	NiMH	LiPo	Lead-Acid
Cost per mAh	Medium	Medium	Low
Weight per mAh	Medium	Low	High
Self-discharge	High (flat in 2–3 months)	Low (6% per month)	Low (4% per month)
Handling of full charge/ discharge cycles	Good	Good	Good
Handling of shallow discharge/charge	Medium (regular full discharge prolongs battery life)	Medium (not well-suited to trickle charging)	Good

TABLE 5-4 Characteristics of Different Battery Technologies

Table 5-4 summarizes the features of NiMH, LiPo, and lead–acid battery technologies.

If you want your project to charge a battery in place, then a LiPo or sealed lead–acid battery is probably the best choice. However, if you want the option to remove the battery and/or use single-use batteries, then a AA battery pack is a good compromise between capacity and size.

For ultra-high-power projects, lead–acid batteries, despite being an ancient technology, still perform pretty well, just as long as you don't have to carry them around! They are also easy to charge and are the most robust of the technologies, offering the least chance of fire or explosion.

Charging Batteries (in General)

Certain principals apply no matter what kind of battery you are charging. So read this section before reading those that follow it concerning specific battery types.

C

The letter C is used to denote the capacity of a battery in Ah or mAh. So, when people talk about charging a battery, they often talk about charging at 0.1C or C/10. Charging a battery at 0.1C means charging it at 1/10 of its capacity per hour. For example, if a battery has a capacity of 2000mAh, then charging it at 0.1C means charging it with a constant current of 200mA.

Over-Charging

Most batteries do not respond well to being over-charged. If you keep supplying them with a high charging current, you will damage them. They often also get hot. In the case of LiPo batteries, this can be "hot" in a fiery sort of way.

For this reason, chargers often charge at a low rate (called trickle charging), so that the low current will not damage the battery. Clearly this makes charging slow. Or, they will use a timer or other circuitry to detect when a battery is full and either stop charging altogether, or switch to trickle charging, which keeps the battery topped up until you are ready to use it.

With some kinds of battery, notably LiPos and the lead–acid variety, if you charge the battery with a constant voltage, then as the battery becomes charged, its voltage rises to match the charging voltage and the current naturally levels off.

Many LiPo batteries now come with a little built-in chip that prevents over-charging automatically. Always look for batteries with such protection.

Over-Discharging

You are probably starting to get the impression that rechargeable batteries are fussy. If so, you're right. Most types of battery are equally unhappy if you over-discharge them and let them go completely flat.

Battery Life

Anyone with a laptop more than a few years old will notice that the capacity of the battery gradually decreases until the laptop only works when plugged in, since the battery has become completely useless. Rechargeable batteries (whatever the technology) can only be recharged a few hundred (perhaps 500) times before needing to be replaced.

Many manufacturers of consumer electronics now build the battery into the device in such a way that it is not "user serviceable," with the rationale that the life of the battery is probably longer than the attention span of the consumer.

How to Charge a NiMH Battery

If you are going to remove your batteries to charge them, this section is pretty trivial. You take them out and put them in a commercial NiMH battery charger that will charge them until they are full and then stop. You can then put them back into your project and you are done.

If, on the other hand, you want to leave the batteries in place while you charge them, then you need to understand a little more about the best way to charge your NiMH batteries.

Simple Charging

The easiest way to charge a NiMH battery pack is to trickle charge it, limiting the current with a resistor. Figure 5-2 shows the schematic for charging a battery pack of four NiMH batteries using a 12V DC adaptor like the one we used back in Chapter 1 to make our fume extractor.

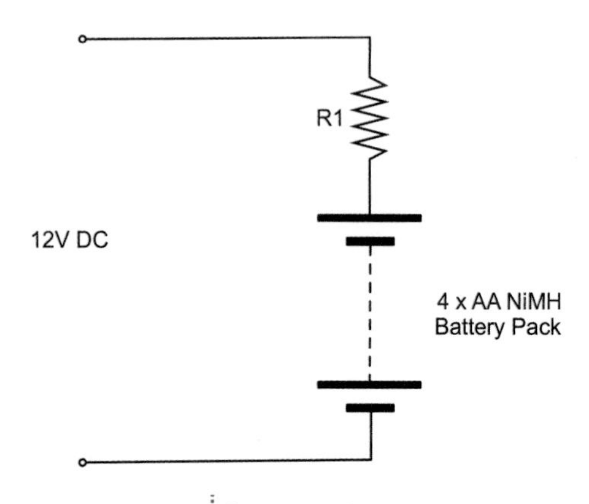

FIGURE 5-2 Schematic for trickle charging a NiMH battery pack

To calculate the value of R1, we first have to decide what current we want to charge our battery with. Generally, a NiMH battery can be safely trickle charged with less than 0.1C indefinitely. If the AA batteries we have each hold a C of 2000mAh, then we can charge them at up to 200mA. To be on the safe side, and if we planned to allow the batteries to "trickle" charge most of the time—for, say, a battery backup project—I would probably use a lower current of 0.05C or more conveniently C/20, which is 100mA.

Typically, the charge time for NiMH batteries is about 3C times the charging current, so at 100mA, we could expect our batteries to take 3 × 2000mAh / 100mA = 60 hours.

Back to calculating R1. When the batteries are discharged, each will be at a voltage of about 1.0V, so the voltage across the resistor will be 12V – 4V = 8V.

Using Ohm's law, R = V / I = 8V / 0.1A = 80Ω.

Let's be conservative and choose the convenient resistor value of 100Ω. Feeding this back in, the actual current will be I = V / R = 8V / 100Ω = 80mA.

When the batteries are fully charged, their voltage will rise to about 1.3V so the current will reduce to: I = V / R = (12V – 1.3V × 4) / 100Ω = 68mA.

That all sounds just fine, our 100Ω will be great. Now we just need to find out what maximum power rating we need for R1.

P = I V = 0.08A × 8 = 0.64W = 640 mW

So, we should probably use a 1-W resistor.

Fast Charging

If you want to charge the batteries faster than that, then it is probably best to use a commercial charger, which will monitor the batteries and turn itself off or reduce the charge to a trickle when the batteries are full.

How to Charge a Sealed Lead–Acid Battery

These batteries are the least delicate of the battery types and could easily be trickle charged using the same approach as for NiMH batteries.

Charging with a Variable Power Supply

However, if you want to charge them faster, then it is best to charge them with a fixed voltage, with some current limiting (a resistor again). For a 12V battery (halve this for a 6V battery) until a discharged battery gets to around 14.4V, you can charge it with almost as much current as your power supply can take. It's only when it gets to this voltage that you need to slow down the charging to a trickle to prevent the battery from getting hot.

The reason we need to limit the current when the battery first starts to charge is that even if the battery doesn't get hot, the wires to it might get hot and whatever is supplying the voltage will only be able to supply a certain amount of current.

Figure 5-3 shows an adjustable power supply. Once you get into electronics, this is one of the first pieces of test equipment you should buy. You can use it in place of batteries while you are working on a project, and also use it to charge up pretty much any type of rechargeable battery.

A variable power supply lets you set both an output voltage and a maximum current. So the power supply will try and supply the specified voltage until the current limit is reached, at which point, the voltage will drop until the current falls back below the set current.

Figure 5-3a shows the power supply set to 14.4V and we have attached the power leads to an empty 12V 1.3Ah sealed lead–acid battery. We will start by adjusting the current setting of the power

(a)

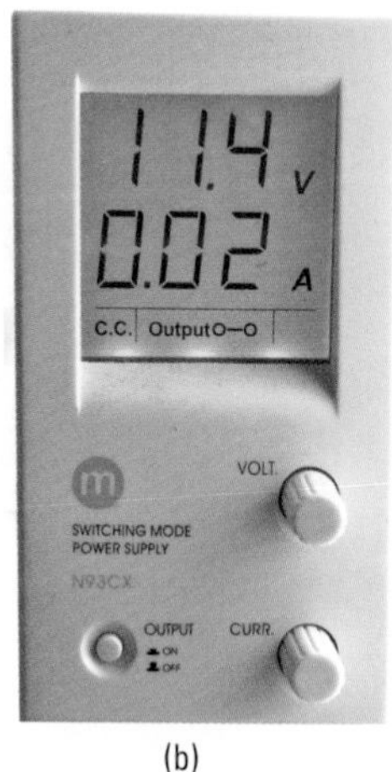

(b)

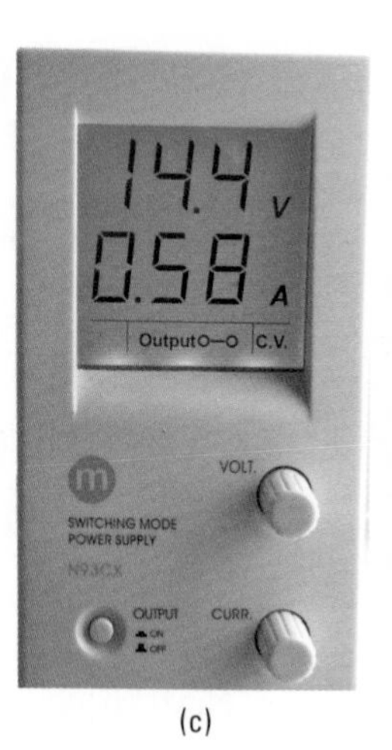

(c)

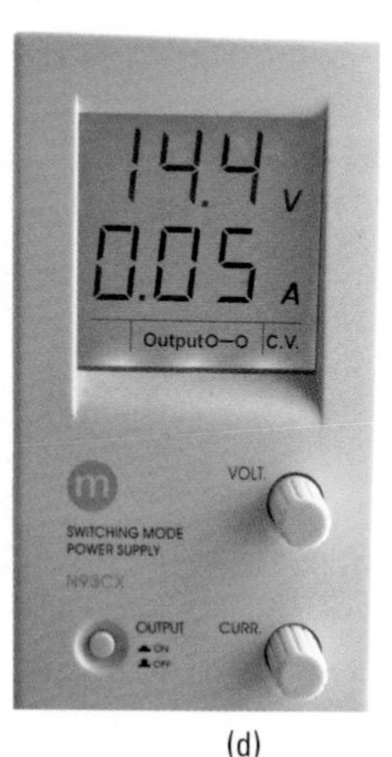

(d)

FIGURE 5-3 Using a variable power supply to charge a lead–acid battery

supply to minimum, so as to prevent any nasty surprises. The voltage immediately drops to 11.4V (Figure 5-3b), so we can gradually increase the maximum current. In actual fact, even with no current limiting (turning the current knob to maximum), the current only rose to 580mA and the voltage increases to 14.4V (Figure 5-3c). After about two hours, the current has dropped to just 200mA, indicating that our battery is getting full. Finally, after four hours, the current is just 50mA and the battery is now fully charged (Figure 5-3d).

How to Charge a LiPo Battery

The technique we have just used on a lead–acid battery using a variable power supply will work just as well on a LiPo battery if we adjust the voltage and current accordingly.

For a LiPo cell, the voltage should be set to 4.2V and the current limited (usually to 0.5A) for a smallish cell, but currents up to C are sometimes used in radio-controlled vehicles.

However, unlike lead–acid and NiMH batteries, you cannot put a number of cells in series and charge the whole lot as one battery. Instead, you have to charge them separately, or use a "balanced charger" that monitors the voltage at each cell separately and controls the power to each.

FIGURE 5-4 SparkFun and generic LiPo chargers

The safest and most reliable way to charge a LiPo is to use one of the chips that exist just for that purpose. These chips are cheap, but generally only available as surface-mounted components. However, there are plenty of ready-made modules available, many of which use the MCP73831 IC. Figure 5-4 shows two of these—one from SparkFun (see the Appendix, M16) and one for just a few dollars from eBay.

Both are used in the same manner. They will charge a single LiPo cell (3.7V) from a USB input of 5V. The SparkFun board has space on the PCB for two other connectors, one to which the battery is connected and the other for a second connection to the battery—the intention is that you connect the electronics that will use the battery to the second socket. The sockets can be either JST connectors as are often found on the end of the leads of a LiPo batter, or just screw terminals. The SparkFun module allows you to select the charging current, using a connection pad.

The generic module on the right has a fixed charge rate of 500mA and just a single pair of connections for the battery.

It is not a good idea to trickle charge a LiPo. If you want to keep them topped up, for say a battery backup solution, then leave them attached to the charger.

Hacking a Cell Phone Battery

Most of us have a cell phone or two languishing in a drawer somewhere, and one of the useful components that can usually be scavenged (assuming it's not the reason the phone is in the drawer) is the battery. The power supply is another useful item.

Figure 5-5a shows a fairly typical vintage cell phone battery. The battery is 3.7V (a single cell) and is 1600mAh (pretty good).

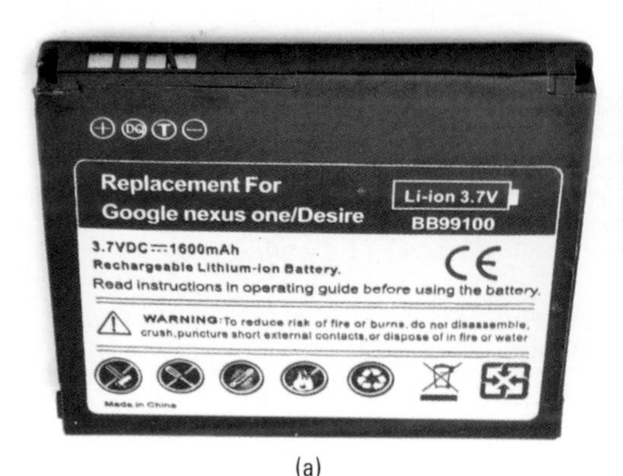

(a)

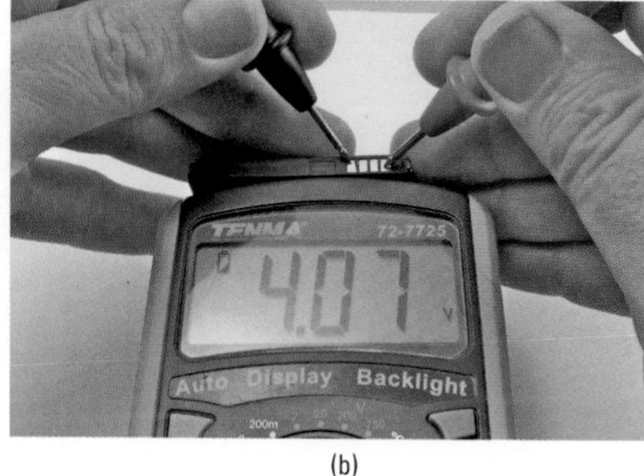

(b)

(c)

Figure 5-5 Hacking a cell phone battery

Cell phone batteries normally have more than just the usual two connections for positive and negative. So the first task must be to identify the connections on the battery.

To identify the positive and negative connections to the cell, just put your multimeter into the 20V DC range and test each combination of pairs until you get the meter to read something over 3.5V, depending on how well charged the battery is (Figure 5-5b).

The batteries often have gold-plated contacts that make them very easy to solder leads to. Once they have leads attached, you can use a charger like the one described in the previous section. Figure 5-5c shows the SparkFun charger module being used for just that purpose.

Caution When using a LiPo battery, remember that if you discharge them too far (below about 3V per cell), you can permanently damage them. Most new LiPo batteries will include an automatic cut-off circuit, built into the battery package, to prevent over-discharging, but this may not be the case for a scavenged battery.

Controlling the Voltage from a Battery

The thing with batteries is that even though they may say 1.5V, 3.7V, or 9V on the package, their voltage will drop as they discharge—often by quite a high percentage.

For example, a 1.5V alkaline AA battery when brand new will be about 1.5V and will quickly fall to about 1.3V under load but still deliver useful amounts of power down to about 1V. This means that in a pack of four AA batteries, the voltage could be anything between 6V and 4V. Most types of battery, whether single-use or rechargeable, exhibit a similar voltage drop.

This may not matter much; it just depends on what the battery is powering. If it is powering a motor or an LED, then the motor will just go a bit more slowly, or the LED will be a little dimmer as the battery discharges. However, some ICs have a very narrow voltage tolerance. There are ICs designed to work at 3.3V that specify a maximum working voltage of 3.6V. Similarly, if the voltage drops too low, the device will also stop working.

In fact, many digital chips such as microcontrollers are designed to work at a standard voltage of 3.3V or 5V.

To ensure a steady voltage, we need to use something called a voltage regulator. Fortunately for us, voltage regulators come as convenient three-pin, low-cost chips that are very easy to use. In fact, the packages just look like transistors, and the bigger the package, the more current they can control.

Figure 5-6 shows how you would use the most common of voltage regulators, called the 7805.

Using just a voltage regulator IC and two capacitors, any input voltage between 7V and 25V can be regulated to a constant 5V. The capacitors provide little reservoirs of charge that keep the regulator IC operating in a stable manner.

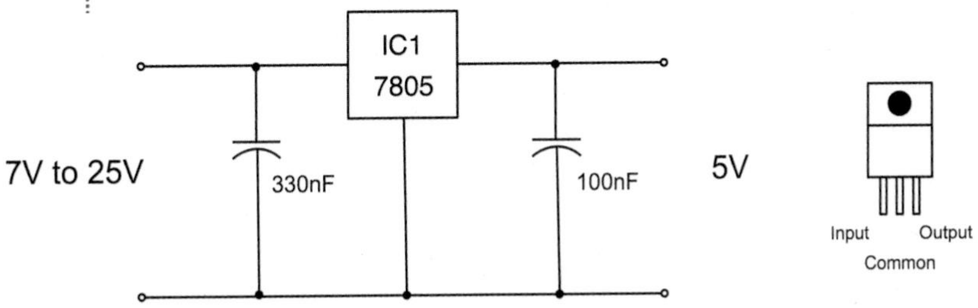

FIGURE 5-6 A voltage regulator schematic

IC1
7805
7V to 25V
330nF
100nF
5V
Input
Output
Common

In the following experiment with a 7805, we will omit the capacitors, as the supply voltage is a steady 9V battery and the load on the output is just a resistor (Figure 5-7).

The capacitors become much more necessary when the load varies (in other words, in the amount of current that it draws), something that is true of most circuits.

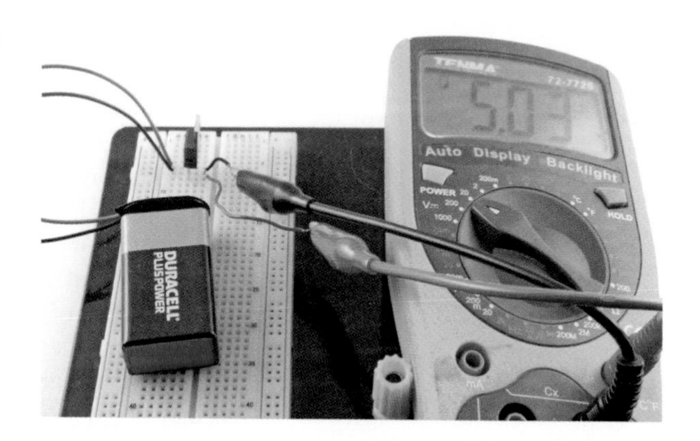

FIGURE 5-7 Experimenting with the 7805

You Will Need

Quantity	Names	Item	Appendix Code
1		Solderless breadboard	T5
1	IC1	7805 voltage regulator	K1, S4
1		Battery clip	H2
1		9V PP3 battery	

Wire up the breadboard as shown in Figure 5-8.

FIGURE 5-8 The 7805 breadboard layout

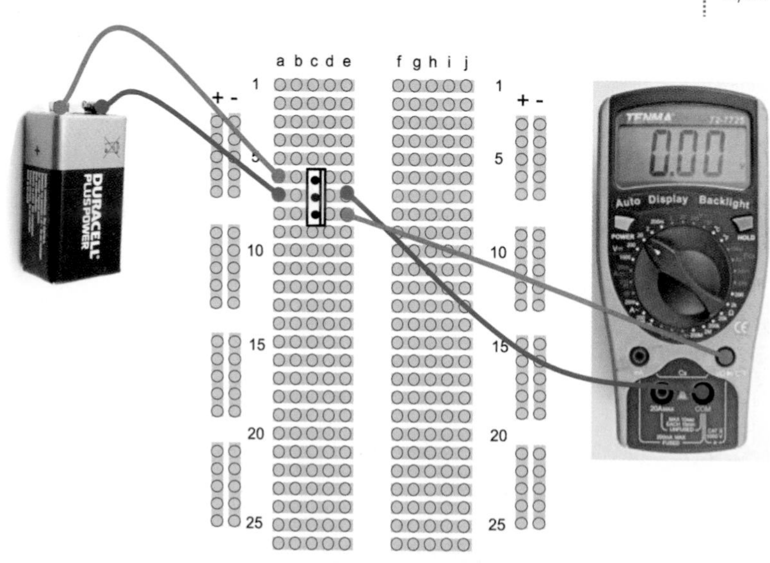

Breadboard

With the battery connected, the multimeter should display a voltage of close to 5V.

Although 5V is a very common voltage, there are voltage regulators for most common voltages, as well as the LM317 voltage regulator that we discussed in Chapter 4, that as well as providing constant current, can also be configured as a voltage regulator.

Table 5-5 lists some common voltage regulators that provide different output voltages and different current handling capabilities.

Output Voltage	100mA	1-2A
3.3V	78L33	LF33CV
5V	78L05	7805 (App A S4) (7–25V in)
9V	78L09	7809
12V	78L12	7812

TABLE 5-5 Voltage Regulators

Boosting Voltage

The voltage regulator ICs in the section titled "Controlling the Voltage from a Battery" only work if the input voltage is greater than the output voltage. In fact, it normally has to be a couple of volts higher, but some more expensive voltage regulators called LDO (low drop out) regulators are available that only require about half a volt more on the input than the output.

Sometimes, however—and cellular phones are a good example of this—it is very convenient to use a single-cell LiPo battery of 3.7V when we require a higher voltage (often 5V) for the circuit.

In these situations, you can employ a very useful circuit called a buck-boost converter. These use an IC and a small inductor (coil of wire) and by applying pulses to the inductor, produce a higher voltage. Actually, it's more complex than that, but you get the idea.

Buck-boost converters are readily available as modules on well-known auction sites. You can find 1A adjustable modules that will provide an adjustable output of 5V to 25V from 3.7V for a few dollars. Try searching for "Boost Step-Up 3.7V." The main module suppliers also provide such boards for around USD 5.

SparkFun sells an interesting module (see the Appendix, M17) that combines a LiPo battery charger with a buck-boost, so you can both charge your LiPo from an external USB 5V input and use the 3.7V LiPo cell to provide an output of 5V using the buck-boost (Figure 5-9).

This actually takes all the difficulties away from using a LiPo in a situation where you want to charge the LiPo battery in situ. Your 5V microcontroller circuit or whatever you are using is just attached to the VCC and GND connections, the battery is clipped into the socket, and to charge the device, you just plug in a USB cable.

FIGURE 5-9 Combined LiPo charger and booster

Calculating How Long a Battery Will Last

We have already touched on the capacity of a battery—that is, the number of mAh it can supply. However, other factors come into play that we should think about when deciding if the batteries we are considering for a project are going to last long enough.

It's really just a matter of common sense, but nevertheless it's easy to make false assumptions about what you need.

As an example, I recently built an automated door for my chicken house. It opens at dawn and closes when it gets dark. It uses an electric motor, and electric motors use a lot of current, so I needed to decide what kind of batteries to use. My first thought was to use big D cells or a lead–acid battery. But when it came to do the math, I found this wasn't really necessary.

Although the motor uses 1A each time it is in operation, it is only in operation twice a day, and each time only for about three seconds. I measured the control circuit as using 1mA all the time. So, let's work out how many mAh the control circuit and motors each use in a day, and then see how many days various types of battery will last.

Let's start with the motors:

$$1\text{A} \times 3 \text{ seconds} \times 2 = 6\text{As} = 6/3600\text{Ah} = 0.0016 \text{ Ah} = 1.6\text{mAh per day}$$

On the other hand, the controller, which I had assumed was the low power part of the project would consume:

$$1\text{mA} \times 24 \text{ hours} = 24\text{mAh per day}$$

This means we can pretty much ignore the power consumed by the motor since it is less than a tenth of the juice required by the controller. Let's say the total requirement is 25mAh/day.

AA batteries are typically 3000mAh, so if we powered the project from AA batteries, we could expect them to last 3000mAh / 25mAh per day = 120 days.

So we do not really need to look much further, AAs will be fine. In the end, I used solar power for this project, which we will visit again in the section titled "How to Use Solar Cells" later in this chapter.

How to Design for Battery Backup

Replacing batteries is a nuisance, and expensive, so it is often cheaper and more convenient to power things from a wall-wart power supply. However, this brings its own disadvantages:

- The device is now tethered to a wire.
- If the household electricity fails, the device will stop working.

The best of both worlds can be achieved by arranging for automatic battery backup of a device that is powered by your household electricity. So, both batteries and a power supply are used, but the batteries are only used if the power supply is not available.

Diodes

What we do not want to happen is for both the batteries and the voltage from the power supply to conflict with each other when both are available. For instance, if the power supply is at a higher voltage than the batteries, it would charge them. But without anything to limit the current, this could be disastrous, even if the batteries were of the rechargeable sort.

Figure 5-10 shows the basic schematic for this. The power supply always needs to be a higher voltage than the battery, so in this case it is 12V and the battery 9V. The schematic also assumes that the battery backup is being used to drive a light bulb.

Recall that diodes act rather like one-way valves. They only allow current to flow in the direction of the arrow. So, let's look at the three possible cases of how power could be supplied here. This is simply the power supply: just batteries and both the battery and power supply (Figure 5-11).

FIGURE 5-10 Battery backup schematic

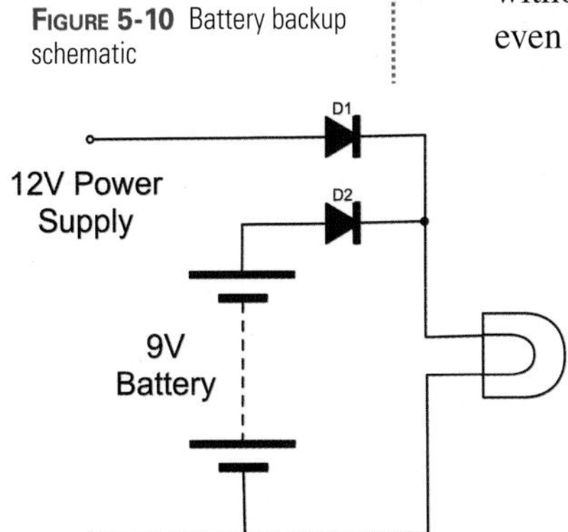

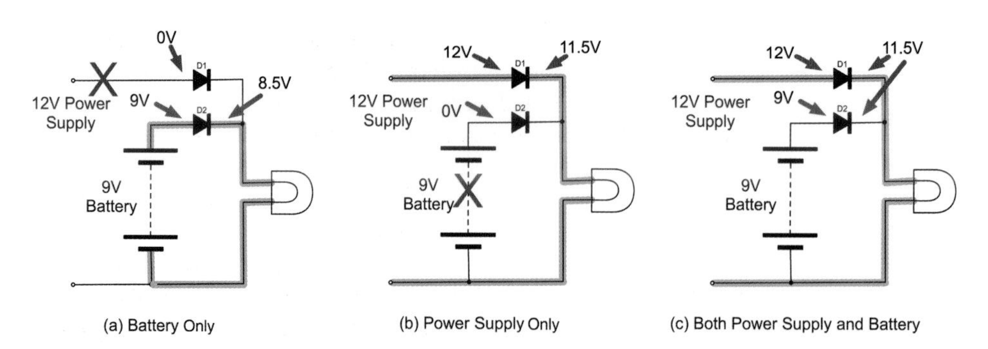

FIGURE 5-11 Diodes for battery backup

Just the Battery

If only the battery has a voltage greater than zero (in other words, the power supply is not plugged in), then the situation is as shown in Figure 5-11a. The 9V from the battery will be at the anode of D2, and the cathode of D2 will be pulled toward ground by the load of the light bulb. This will cause D2 to be forward-biased and conduct the current through the light bulb. A forward-biased diode will have an almost constant voltage of 0.5V across it, which is why we can say that the voltage after the diode is 8.5V.

On the other hand, D1 will have a higher voltage (8.5V) on its cathode (right-hand side in the diagram) than its anode (0V), so no current will flow through D1.

Just the Power Supply

If just the power supply is connected (Figure 5-11b), then the role of the diodes is reversed and now the current flows through D1 to the light bulb.

Both the Power Supply and the Battery

Figure 5-11c shows the situation where both the power supply and the battery are connected. The 12V of the power supply will ensure that the cathode of D2 is at 11.5V. Since the anode of D2 is at 9V from the battery, the diode will remain reverse-biased and no current will flow through it.

Trickle Charging

As we already have a battery and a power supply, we have most of the ingredients we need to charge the battery. We could for example use six AA rechargeable batteries in a battery box

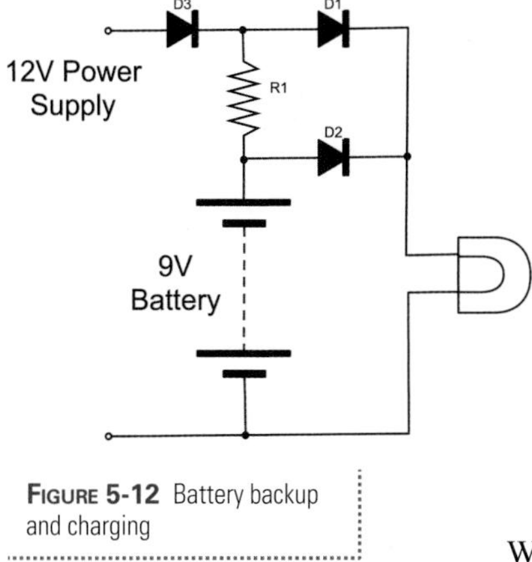

FIGURE 5-12 Battery backup and charging

and arrange to charge them at C/20 (assuming C = 2000mAh) or 100mA from the power supply.

That way, the batteries would always be charged, and provide light whenever the power failed. Figure 5-12 shows the schematic for this.

You may not have been expecting the extra diode D3. This is really just to account for the fact that we do not know exactly how the power supply is designed, so we do not know what would happen if the battery was connected to its output (via R1) when it was turned off. This may discharge the battery or damage the power supply. The diode D3 just protects it and makes sure no current can flow back into it.

We want a charging current of 100mA to flow through R1, and we know that when both the power supply and battery are connected, there will be a voltage across R1 of 12V – 0.5V – 9V or 2.5V. So, using Ohm's law, the value of the resistor should be:

$$R = V / I = 2.5 / 0.1A = 25\Omega$$

The nearest standard value to this is probably 27Ω.

Its power requirement: $P = V^2 / R = 2.5^2 / 27 = 0.23W$

This means a standard half- or quarter-watt resistor will be fine.

How to Use Solar Cells

On the face of it, solar cells seem like the perfect power source. They convert light into electricity, and so in theory you need never change a battery or be plugged into a wall outlet again!

However, as always, the reality is not quite so simple. Solar cells, unless they are very large, produce fairly small amounts of electricity and so are most suited to low-power devices and projects that are outdoors away from household electricity.

If you are thinking of trying a solar project that will be installed indoors, unless it will be sited against a south-facing window, I really wouldn't try it. Solar cells do not require direct sunlight, but to produce any useful amounts of electricity, they really need a good unobstructed view of the sky.

Two solar projects I have developed are a solar-powered radio (the solar panel is as big as the radio and, yes, it needs to be next to the window), and a solar-operated chicken house door. If you are lucky enough to live somewhere sunny, then solar power is obviously a lot easier.

Figure 5-13 shows a typical solar panel. This one was scavenged from a security light installation. It is about six inches by four inches and has a swivel mount that allows it to be angled toward the sun. It is the panel I used for the chicken house door.

Figure 5-13 A solar panel

Projects that use a solar panel to provide power nearly always also use a rechargeable battery. So the panel charges the battery and the project draws its power from the battery.

Small solar cells generally only produce around half a volt, so they are usually combined into panels of many cells that increase the voltage to a level that is high enough to charge a battery.

The voltage you find on a solar panel normally refers to the voltage of battery that the solar panel is capable of charging. So, it is quite common to find 6V or 12V solar panels. When you measure the voltage from these in bright sunlight, the reading will be much higher, possibly 20V for a 12V panel. But, under the load of charging a battery, this drops rapidly.

Testing a Solar Panel

A solar panel will have a certain number of watts and a nominal voltage specified for it. These tend to be for ideal conditions, so when I get a solar panel that I want to use in a project, I like to test it to find what it is really capable of. Without knowing how much power it can provide in a real situation where it's installed, it is hard to make safe assumptions about battery capacities and how low you need to keep the current consumption.

When testing out a solar panel, you should use a resistor as a "dummy load," and then try out the solar panel in various locations and levels of brightness, measuring the voltage across the resistor. From this, you can calculate the current being provided by the panel.

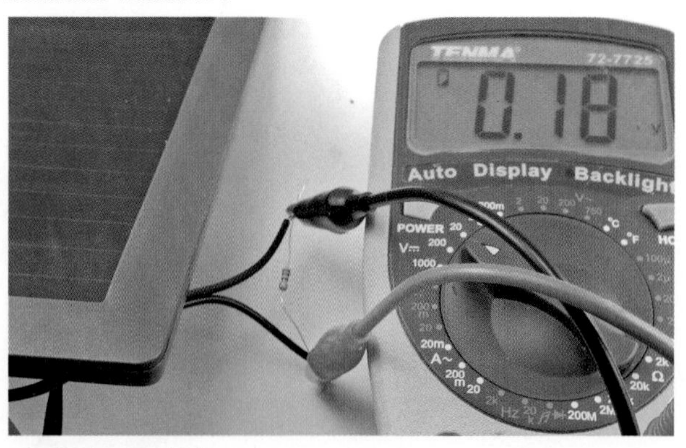

Figure 5-14 Testing a solar panel

Figure 5-14 shows such an arrangement for my "chicken house" solar panel. The meter is showing just 0.18V with

Lead–acid batteries are still a very popular choice for trickle charging from solar. This is mainly because they are very forgiving of gentle over-charging and have a lower self-discharge rate than, say, NiMH batteries.

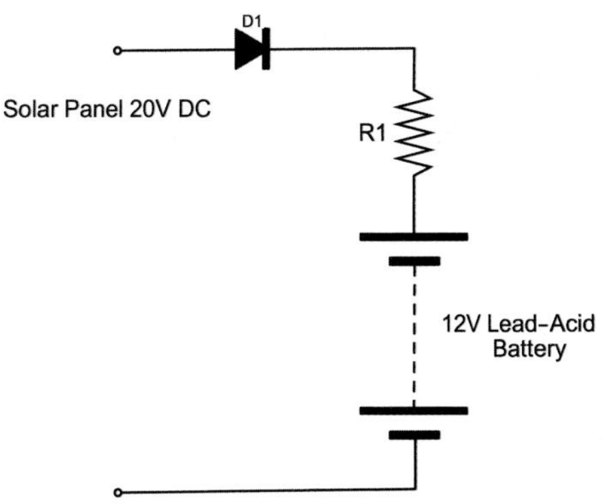

FIGURE 5-16 Schematic for solar trickle charging

Minimizing Power Consumption

When planning solar power for some small outdoor project, you need to make sure the solar panel charging the battery can keep up with demand.

If you live in southern California, the design for using solar panels is pretty easy. You can count on quite a lot of sun all year long. However, if you live a long way from the equator, say, in a maritime climate where it's often quite dull during the day, then you will have short winter days. You may get weeks of dull weather with short days. If your system is to work all year long, you either need to have a large battery that will last for a few weeks of dull weather, or use a larger solar panel.

The sums are pretty easy. There are mAh going into the battery from the solar panel, and mAh coming out for the device it is powering. The device might be running all the time, but the solar panel is only active half the time (daylight). So, you need to work out what you think the worst case for solar input might be for a week or two, and then design it accordingly.

It will probably be easier and cheaper to put your efforts into minimizing the current consumed by the system rather than increasing the size of the solar panel and battery.

Summary

In this chapter, we have learned about how to power our projects. In the next chapter, you will learn how to use the very popular Arduino microcontroller board.

a 100Ω load resistor inside the light box that I use for my photography. That equates to just 1.8 mA.

I find a spreadsheet a useful way of recording how the solar panel performs. Figure 5-15 shows an excerpt from the spreadsheet, complete with graph. You can then file this away until the next time you wish to use a solar panel in a project.

The spreadsheet can be downloaded from www.hackingelectronics.com, but there is really nothing complex about the math.

As you can see, the solar panel produces only 1 or 2mA indoors even under bright artificial lighting. The results outdoors with a clear view of the sky are better, but it really only produces quite high power in direct sunlight.

Trickle Charging with a Solar Panel

Since the solar panels produce a reasonable voltage, even in relatively low light conditions, they can easily be used to trickle charge a battery. However, you should always use a diode to protect the solar panel from the situation where the battery is at a higher voltage than the panel (say at night), since such a reverse flow will damage the solar panel.

A typical simple trickle charge schematic is shown in Figure 5-16.

FIGURE 5-15 Solar panel data

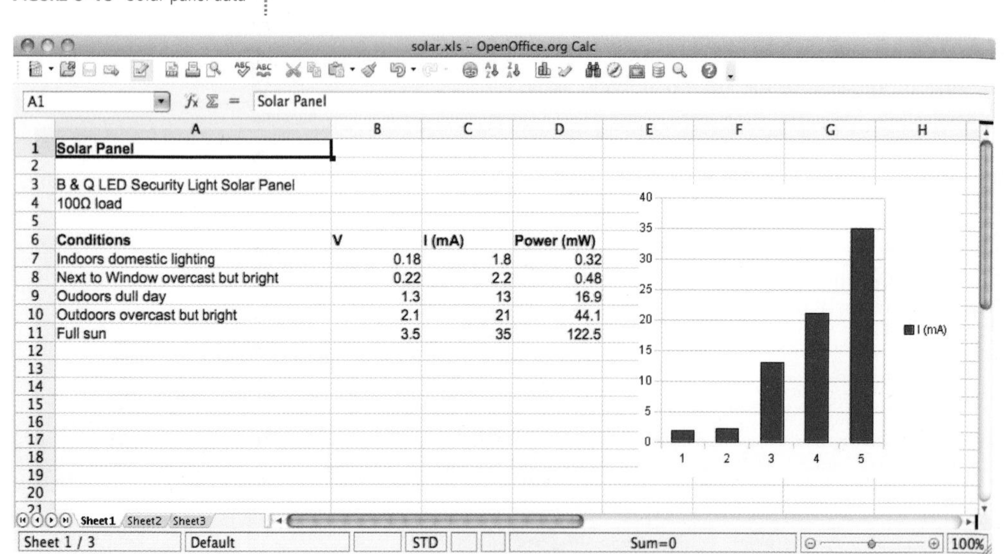

solar.xls - OpenOffice.org Calc

Solar Panel

B & Q LED Security Light Solar Panel

100Ω load

Conditions	V	I (mA)	Power (mW)
Indoors domestic lighting	0.18	1.8	0.32
Next to Window overcast but bright	0.22	2.2	0.48
Oudoors dull day	1.3	13	16.9
Outdoors overcast but bright	2.1	21	44.1
Full sun	3.5	35	122.5

6

Hacking Arduino

Microcontrollers are essentially low-powered computers on a chip. They have input/output pins to which you can attach electronics so the microcontroller can, well, control things. Utilizing a microcontroller used to be quite a complex process, largely because the microcontroller needed to be programmed. This was often done in assembler or complex C. But there was a lot to learn before you could do anything useful. Because of this, it discouraged their use in casual projects where you just wanted to hack something together.

Enter the Arduino (Figure 6-1). The Arduino is a simple-to-use, low-cost, readymade board that lets you use a microcontroller in your projects with a minimum of fuss.

The Arduino sells in vast quantities and has become the platform of choice for makers and hackers in need of microcontrollers.

The popularity of Arduino is due to many factors, including its:

- Low cost
- Open-source hardware design
- Easy-to-use integrated development environment (IDE) to program it with
- Plug-in shields that add features like displays and motor drivers that clip onto the top of the Arduino

All the programs for the Arduino used in this and later chapters are available for download from the book's accompanying web site (www.hackingelectronics.com).

The examples in this book are designed to work with both the Arduino Uno and the Arduino Leonardo. However, two of the projects, in the sections "How to Type Passwords Automatically" and "How to Make a USB Music Controller" (see Chapter 9), only work with Arduino Leonardo.

The Leonardo is the newer board. You may have some compatibility problems with this board and some Arduino shields. This includes any Ethernet Shield prior to the R3 Ethernet Shield. So, if you have an older Ethernet Shield, then the section "How to Control a Relay from a Web Page" will work on an Arduino Uno, but not a Leonardo, unless you have an R3 shield.

FIGURE 6-1 An Arduino Uno board

How to Set up Arduino (and Blink an LED)

To be able to program an Arduino, we first have to install the Arduino integrated development environment (IDE) on our computer. Arduino is available for Windows, Mac, and Linux.

You Will Need

Quantity	Item	Appendix Code
1	Arduino Uno/Leonardo	M2/M21
1	USB lead; Type B for Uno, Micro USB for Leonardo	

Setting Up Arduino

The first step is to download the software for your computer type from the official Arduino web site here: http://arduino.cc/en/Main/Software.

Once this is downloaded, you can find detailed instructions for the installation of each platform here: http://arduino.cc/en/Guide/HomePage.

One of the nice things about the Arduino is that to get started with it, all you need is an Arduino, a computer, and a USB lead to connect the two together. The Arduino can even be powered

FIGURE 6-2 The Arduino, laptop, and chicken

over the USB connection to the computer. Figure 6-2 shows an Arduino Uno (the most common type of Arduino) connected to a laptop running the Arduino IDE.

To prove that the Arduino is working, we will program it to flash an LED that is on the Arduino board labeled "L" and hence known as the "L" LED.

Start by launching the Arduino IDE on your computer. Then, from the File menu (Figure 6-3), select Examples | 01.Basics | Blink.

In an attempt to make programming the Arduino sound less daunting to non-programmers, programs on the Arduino are referred to as "sketches." Before we can send the Blink sketch to your Arduino, we need to tell the Arduino IDE what type of Arduino we are using. The most common type is the Arduino Uno, and in this chapter we will assume that is what you have. So from the Tools | Board menu, select Arduino Uno (Figure 6-4).

As well as selecting the board type, we also need to select the port it is connected to. In Windows, this is easy because it is always COM3 or COM4 (see Figure 6-5). However, on a Mac or Linux, there will generally be more serial

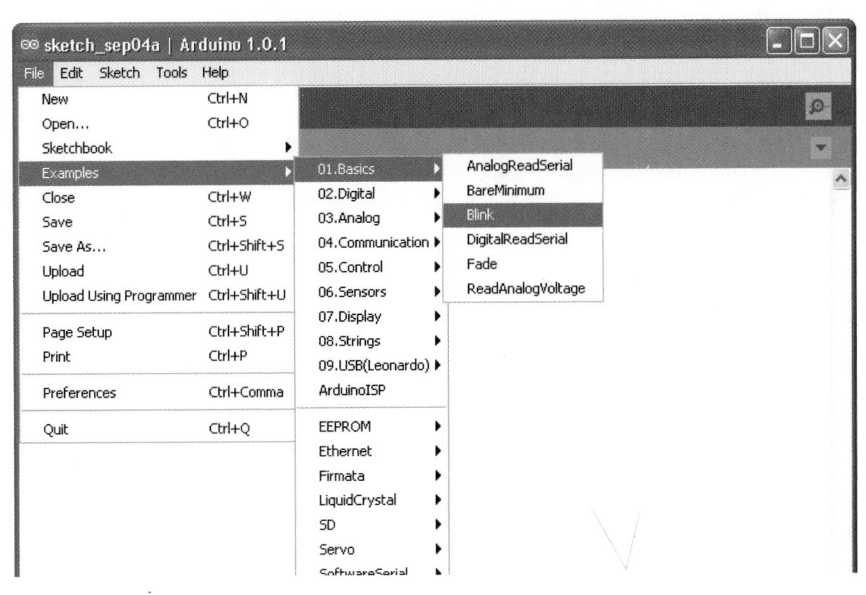

FIGURE 6-3 Loading the "Blink" sketch

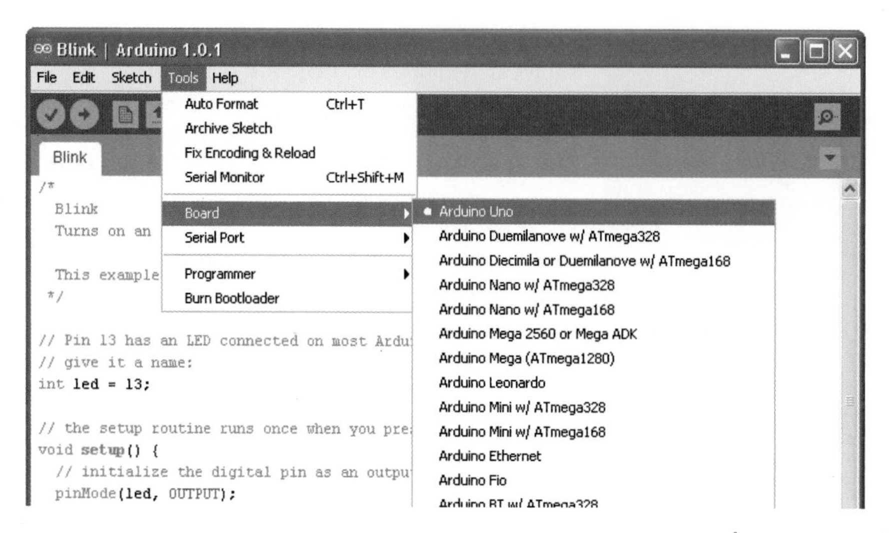

FIGURE 6-4 Selecting the board type

devices. However, the Arduino IDE shows them with the most recently connected first, so your Arduino board should be at the top of the list.

To actually upload the sketch onto the Arduino board, click the Upload button on the tool bar. This is the second button on the toolbar, highlighted in Figure 6-6.

Once, you press the Upload button, a few things should happen. First, a progress bar will appear as the Arduino IDE first compiles the sketch (converts it into a suitable form for uploading). Then, the LEDs on the Arduino—labeled Rx and Tx—should flicker for a while.

Finally, the LED labeled L should start to blink. The Arduino IDE will also show a message that looks something like "Binary sketch size: 1,084 bytes (of a 32,256 byte maximum)." This means

FIGURE 6-5 Selecting the serial port

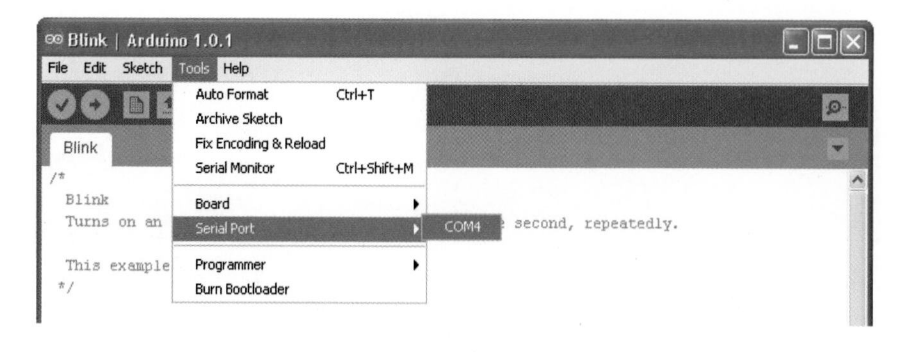

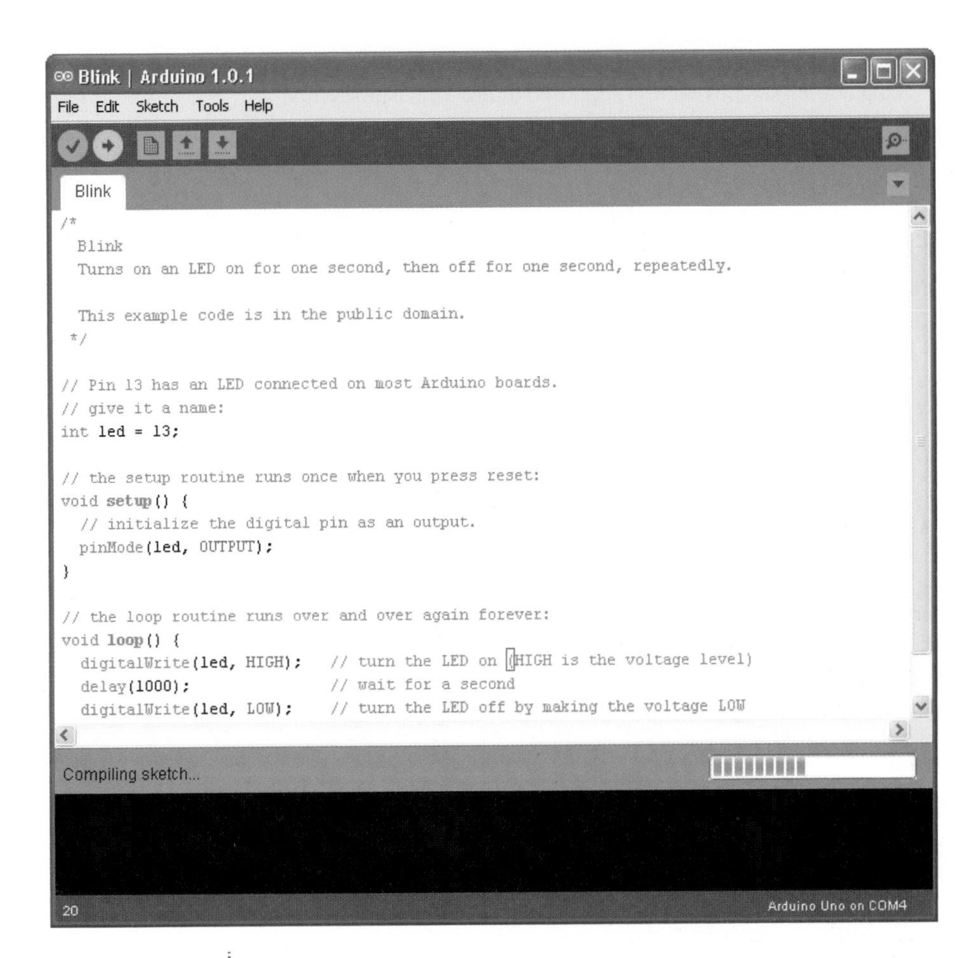

FIGURE 6-6 Uploading the Blink sketch

the sketch has used about 1kb of the 32k of flash memory available for programs on the Arduino.

Note that if you are using a Leonardo, you may have to keep the Reset button depressed until you see the message "Uploading…" in the Arduino software.

Modifying the Blink Sketch

It may be that your Arduino was already blinking when you first plugged it in. That is because the Arduino is often shipped with the Blink sketch already installed.

If this is the case, you might like to prove to yourself that you have actually done something by changing the blink rate. We will now examine the Blink sketch and see how we could change it to make it blink faster.

The first part of the sketch is just a comment, to tell someone looking at the sketch what it is supposed to do. This is not actual program code, and part of the preparation for the code being uploaded is for all such "comments" to be stripped out. Anything between "/*" and "*/" is ignored.

```
/*
  Blink
  Turns on an LED on for one second, then off for one
second, repeatedly.

  This example code is in the public domain.
 */
```

We then have a couple of individual line comments that start with //. Just like the block comments earlier, these are simply to inform us of what is happening. In this case, it helpfully tells us to use pin 13 as the pin we will flash. We have chosen that pin because on an Arduino Uno that pin is connected to the built-in "L" LED.

```
// Pin 13 has an LED connected on most Arduino boards.
// give it a name:
int led = 13;
```

The next part of the sketch is the "setup" function. Every Arduino sketch must have a "setup" function, and this function runs every time the Arduino is reset, either because (as the comment says) the reset button is pressed, or the Arduino is powered up.

```
// the setup routine runs once when you press reset:
void setup() {
  // initialize the digital pin as an output.
  pinMode(led, OUTPUT);
}
```

The structure of this text is a little confusing if you are new to programming. A function is a section of code that has been given a name (in this case, the name is "setup"). For now, simply use the text just cited as a template and understand that it must start with the first line "void setup() {". Afterward, you place each of the commands you want to issue on a line ending with ";", and then mark the end of the function with a "}" symbol.

In this case, the only command we expect the Arduino to carry out is to issue the "pinMode(led, OUTPUT)" command that not unsurprisingly sets that pin to be an output.

Next comes the juicy part of the sketch: the "loop" function.

Like the "setup" function, every Arduino sketch must have a "loop" function. Unlike "setup" which only runs once after a reset, the "loop" function runs continuously. That is, as soon as all its instructions have been done, it starts again.

In the "loop" function, we first turn on the LED by issuing the "digitalWrite (led, HIGH)" instruction. We then pause for a second by using the command "delay(1000)". The value is 1000 for 1000 milliseconds, or one second. We then turn the LED back off again, and delay for another second before the whole process starts over.

```
// the loop routine runs over and over again forever:

void loop() {
  digitalWrite(led, HIGH);    // turn the LED on (HIGH is the voltage level)
  delay(1000);                // wait for a second
  digitalWrite(led, LOW);     // turn the LED off by making the voltage LOW
  delay(1000);                // wait for a second
}
```

To modify this sketch to make the LED blink faster, change both occurrences of 1000 to 200. These changes are both in the "loop" function, so your function will now look like this:

```
void loop() {
  digitalWrite(led, HIGH);    // turn the LED on (HIGH is the voltage level)
  delay(200);               // wait for a second
  digitalWrite(led, LOW);     // turn the LED off by making the voltage LOW
  delay(200);               // wait for a second
}
```

If you try to save the sketch before uploading it, you will be reminded that it is a "read-only" example sketch, but the Arduino IDE will offer you the option to save it as a copy, which you can then modify to your heart's content.

You do not have to do this, of course. You can just upload the sketch unsaved, but if you do decide to save this or any other sketch, you will find that it then appears in the File | Sketchbook menu on the Arduino IDE.

So, either way, click the Upload button again. When the uploading is complete, the Arduino will reset itself and the LED should start to blink much faster.

How to Make an Arduino Control a Relay

The USB connection of an Arduino can be used for more than just programming the Arduino. You can also use it to send data between the Arduino and your computer. If we attach a relay to the Arduino, we could send a command from our computer to turn the relay on and off.

Relays

A relay (Figure 6-7) is an electromechanical switch. It's very old technology, but relays are cheap and very easy to use.

A relay is basically an electromagnet that closes switch contacts. The fact that the coil and the contacts are electrically isolated from one another makes relays great for things like switching home-powered devices on and off from something like an Arduino.

Whereas the coil of a relay is often energized by between 5V and 12V, the switch contacts can control high-power, high-voltage loads. For example, the relay photographed in Figure 6-7 claims a maximum current of 10A at 120V AC (household power) as well as 10A at 24V DC.

Arduino Outputs

Arduino outputs, and for that matter inputs, are referred to as "pins," even though if you look at the connectors along the sides of the Arduino, they are most definitely sockets rather than pins. The name harkens back to the pins on the microcontroller IC at the heart of the Arduino that were connected to the sockets.

Each of these "pins" can be configured to act as either an input or an output. When they are acting as an output, each pin can provide up to 40mA. This is more than enough to light

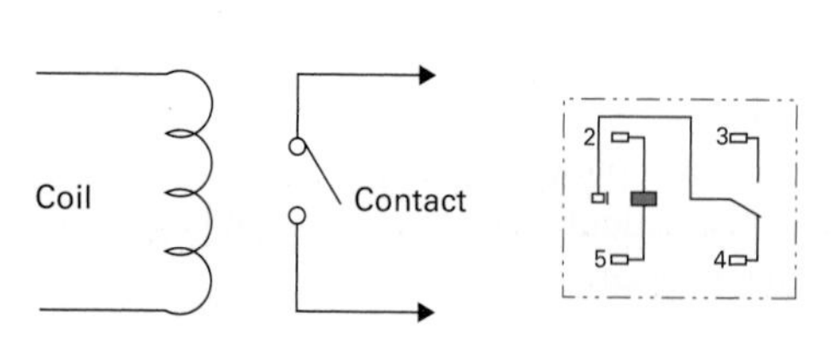

Relay Schemtic

Relay Package

A Relay

FIGURE 6-7 A relay

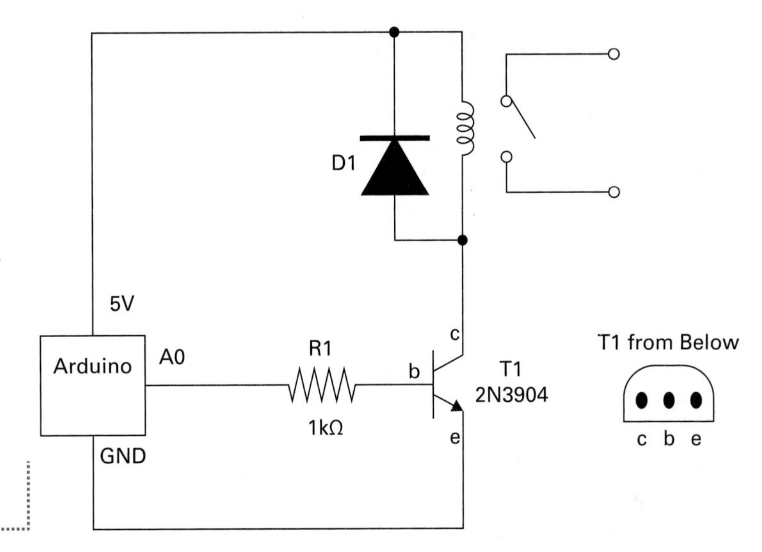

FIGURE 6-8 Schematic diagram of an Arduino-controlled relay

an LED, but not enough to energize a relay coil, which typically requires more like 100mA.

This is a problem we have already discussed. Since we want to use a small current to control a larger one, a good way to do this is by using a transistor.

Figure 6-8 shows a schematic diagram of what we are going to build.

We are using a transistor just like we did when we were controlling an LED. One difference in the schematic is that there is a diode across the relay coil. This is required because when you turn the relay off and the magnetic field in the coil collapses, you get a spike of voltage. The diode prevents this from damaging anything.

We are going to solder the components to the relay and then attach the necessary leads to a header strip that will plug into the Arduino (Figure 6-9). The header strip has 15 pins and spans both of the connector sockets on the side of the Arduino closer to the microcontroller chip. There is a gap between the two connector strips, so one of the header pins will not actually be fitted into a socket.

FIGURE 6-9 The Arduino relay interface

You Will Need

Quantity	Names	Item	Appendix Code
1		Arduino Uno/Leonardo	M2/M21
1		USB lead; Type B for Uno, Micro USB for Leonardo	
1		Transistor 2N3904	K1, S1
1	R1	1kΩ 0.25-W resistor	K2
1	D1	1N4001 diode	K1, S5
1	Relay	5V Relay	H16
1		* Pin header 15-way	K1, H4
1		Two-way screw terminal	H5

* Pin headers are usually supplied in long lengths designed to be snapped into whatever number of connections you need.

Construction

Figure 6-10 shows how the components are attached. First, solder the diode to the relay coil contacts. These are the two pins on the far side of the relay that have three pins more or less in a row. The stripe on the diode should be to the right, as shown in Figure 6-10.

After soldering the diode across the relay coils, bend out the leads of the transistor and position it as shown in Figure 6-10 with the flat side against the relay. Shorten the base (middle) lead of the transistor, shorten the leads on the resistor and attach it to the base lead.

Finally, solder the three leads to the connector strip. The resistor lead should go to the 6th lead from the left, the emitter of the transistor to the 9th from the left, and the diode lead to the 11th from the left.

Before we attach leads to the relay contacts, we can test our work using a multimeter in Continuity mode, so attach the header pins to the Arduino as shown in Figure 6-9 and clip one lead of the multimeter (on Continuity mode) to the middle contact of the relay (in between the diode leads). Attach the other lead of the multimeter to each of the two unconnected contacts on the relay. One will buzz and the other will not. Attach the lead to

FIGURE 6-10 Wiring the relay interface

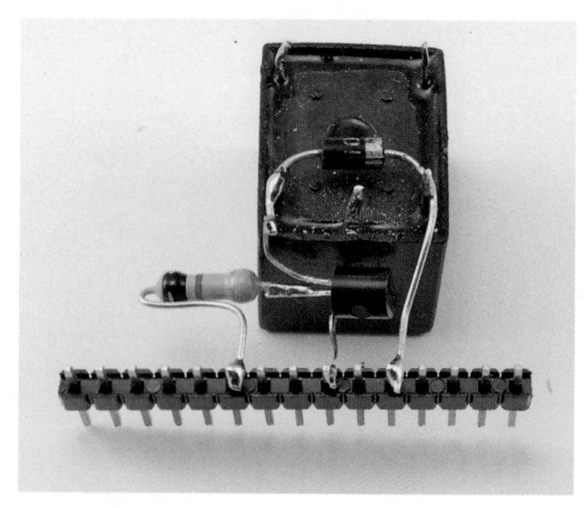

the one that does not cause the multimeter to beep—this is the n.o. (normally open contact).

Load the sketch "relay_test" into Arduino and upload it to the Arduino board. When the Arduino restarts, you should find that every two seconds the relay will flip from being open to being closed.

Software

The sketch for this is much the same as the Blink sketch.

```
// relay_test

int relayPin = A0;

void setup()
{
  pinMode(relayPin, OUTPUT);
}

void loop()
{
  digitalWrite(relayPin, HIGH);
  delay(2000);
  digitalWrite(relayPin, LOW);
  delay(2000);
}
```

The only real difference is that we are using pin A0 rather than pin 13. Arduino has a feature that allows you to use the analog input pins A0 to A5 as digital inputs or outputs as well as analog inputs, but you have to put the letter A in front of them when using them as digital pins.

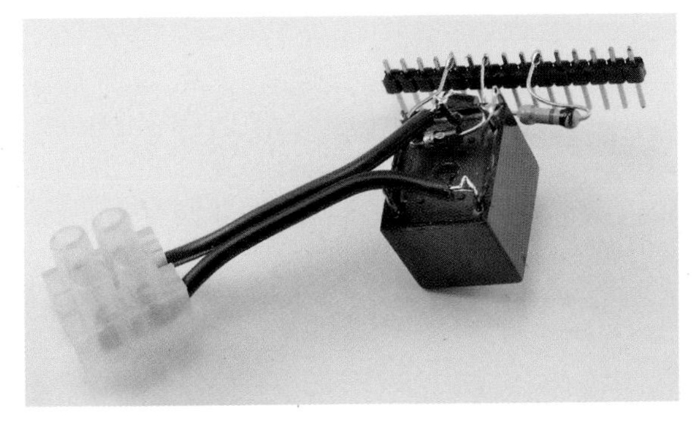

FIGURE 6-11 Attaching leads to the relay contacts

If all is well, then to make it easier to attach things to the relay contacts, we can solder some wires to them and use a two-way screw terminal block (Figure 6-11).

The relay module can be used to control all sorts of things. It can be used to control 110 or 240V home-powered devices, but be very sure you know what you are doing before you attempt that. If you plan to do this, everything must be insulated and the whole project enclosed in a plastic box, especially since touching a "hot" wire can and does kill many people every year.

In the next section, you will hack an electrical toy so it can be turned on and off using the Arduino and relay module you have just built.

How to Hack a Toy for Arduino Control

The great thing about a relay is that it behaves just like a switch. This means that if you have something you want to turn on and off from your Arduino and that item has a switch, then all you need to do is solder some wires to the switch and attach them to the relay. This would allow both the relay and the switch to turn the device on and off. But if you do not want to keep the original switch, it can be removed, as it will be in this case.

The toy that the author chose to hack is a little electric bug (Figure 6-12).

FIGURE 6-12 The hapless electric bug awaiting dissection

You Will Need

As well as the relay module built in the section "How to Make an Arduino Control a Relay," you will also need the following items.

Quantity	Item	Appendix Code
1	Arduino Uno/Leonardo	M2/21
1	USB lead; Type B for Uno, Micro USB for Leonardo	
1	An electric toy (battery-powered) with an on–off switch	
1	Twin multi-core wire	

Construction

Taking the toy apart, you can see the connections to the switch (Figure 6-13a). De-solder the switch and attach wires to the leads that used to go to the switch (Figure 6-13b). You should always put insulating tape around the bare wires to prevent accidental shorts (Figure 6-13c).

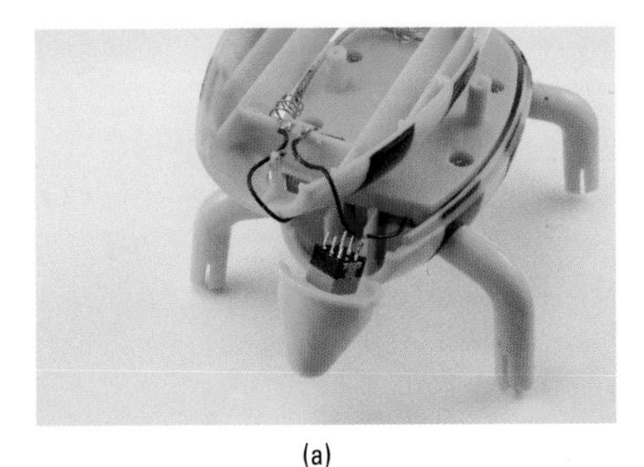

(a)

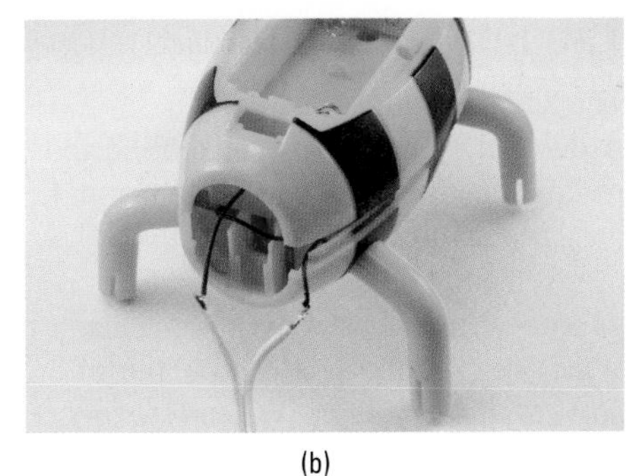

(b)

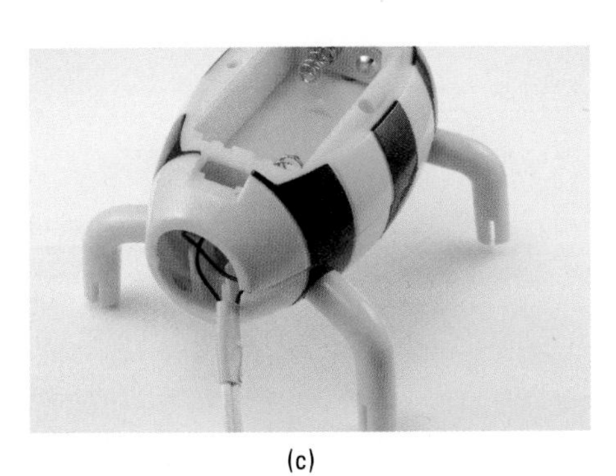

(c)

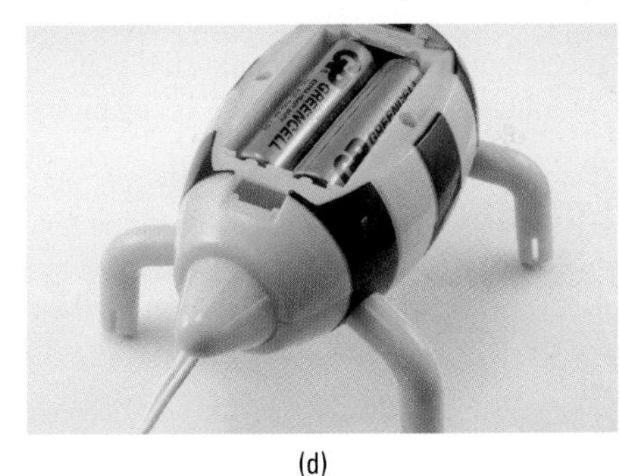

(d)

(e)

Figure 6-13 Hacking the toy

The toy can then be assembled with the wire leaving through a gap in the case (Figure 6-13d). If there is no suitable gap, you will probably need to drill a hole.

Finally, the toy is ready to use, so plug the relay interface into the Arduino and connect the wires to its screw terminals (Figure 6-13e). If the test sketch is still installed, you should find that the toy repeatedly turns on for a couple of seconds, and then turns back off again.

This is okay, but not terribly useful. We will use another sketch that will allow us to send commands to the Arduino from your computer. The sketch is called "relay_remote".

Upload this sketch to the Arduino. Then, open the Serial Monitor by clicking the button on the right-hand side of the Arduino IDE (circled in Figure 6-14).

The Serial Monitor

The Serial Monitor is part of the Arduino IDE that allows you to send and receive data between your computer and the Arduino board (Figure 6-15).

At the top of the Serial Monitor is an area where we can type commands. When we click the Send button, these are sent to the Arduino. We can see any messages that the Arduino has sent in the area below this.

Try this out by typing in the number 1 and clicking Send. This should start your toy. Entering "0" should turn it off again.

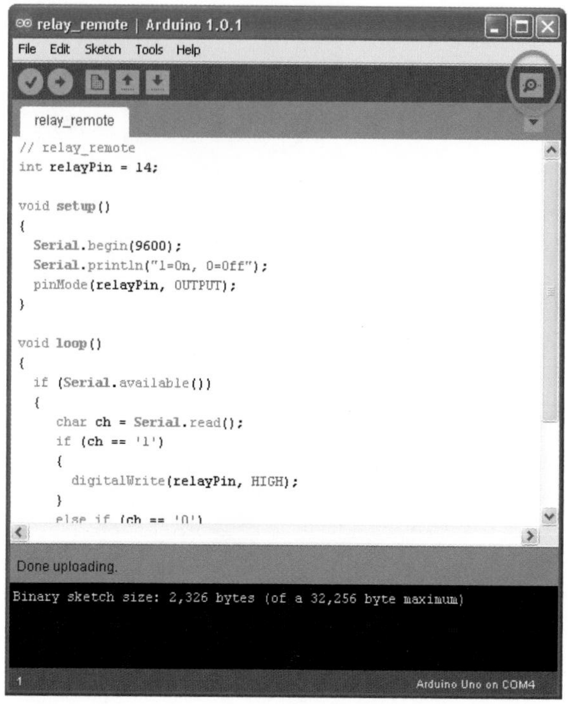

FIGURE 6-14 Opening the Serial Monitor

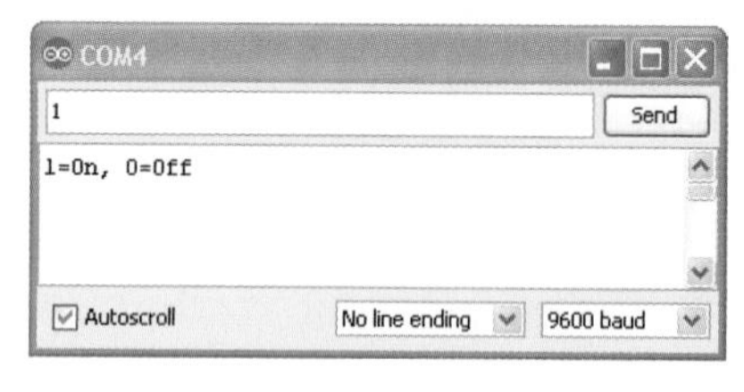

FIGURE 6-15 The Serial Monitor

Software

Let's now look at the sketch.

```
// relay_remote
int relayPin = A0;

void setup()
{
  Serial.begin(9600);
  Serial.println("1=On, 0=Off");
  pinMode(relayPin, OUTPUT);
}

void loop()
{
  if (Serial.available())
  {
     char ch = Serial.read();
     if (ch == '1')
     {
       digitalWrite(relayPin, HIGH);
     }
     else if (ch == '0')
     {
       digitalWrite(relayPin, LOW);
     }
  }
}
```

Notice that the "setup" function now has two new commands in it.

```
Serial.begin(9600);
Serial.println("1=On, 0=Off");
```

The first of these starts serial communications over the serial port at 9600 baud. The second sends the prompt message, so that we know what to do when the Serial Monitor opens.

The "loop" function first uses the function "Serial.available()" to check if there is any communication from the computer waiting to be processed. If there is, then this is read into a character variable.

We then have two if statements. The first checks to see if the character is a "1", and if it is, it turns on the toy. If, on the other hand, the character read is a "0", it turns it off.

We have made a little bit of a leap from our first flashing sketch, and if you need more help understanding how the sketch works, you might consider buying the book *Programming Arduino: Getting Started with Sketches* by this author.

How to Measure Voltage with an Arduino

The pins labeled A0 to A5 on an Arduino are analog inputs. That is, you can use them to measure voltage. To demonstrate this, you will use the variable resistor (trimpot) as a voltage divider connected to A3 (Figure 6-16).

FIGURE 6-16 A variable resistor and Arduino

If you skipped the section on voltage dividers in Chapter 3 titled "How to Use Resistors to Divide a Voltage," you probably should go back and have a quick look now.

You Will Need

To try this example, you will need the following items.

Quantity	Names	Item	Appendix Code
1		Arduino Uno/Leonardo	M2/M21
1		USB lead; Type B for Uno, Micro USB for Leonardo	
1	R1	10kΩ variable resistor	K1, R1

Construction

The construction of this project is very simple. There is no actual soldering involved, we will just push the three pins of the variable resistor into the Arduino sockets A2, A3, and A4. Figure 6-17 shows the schematic diagram for this.

You may be wondering how this will work, since you would normally expect the top of the variable resistor to be connected to 5V and the bottom to GND. Well, since a 10kΩ resistor at 5V will only be allowing 0.5mA of current to flow, we can use pins A2 as digital outputs and set A2 to 0V and A4 to 5V.

Plug the variable resistor into the Arduino so that the middle slider connection is to pin A3 and the pins on either side are connected to A2 and A4.

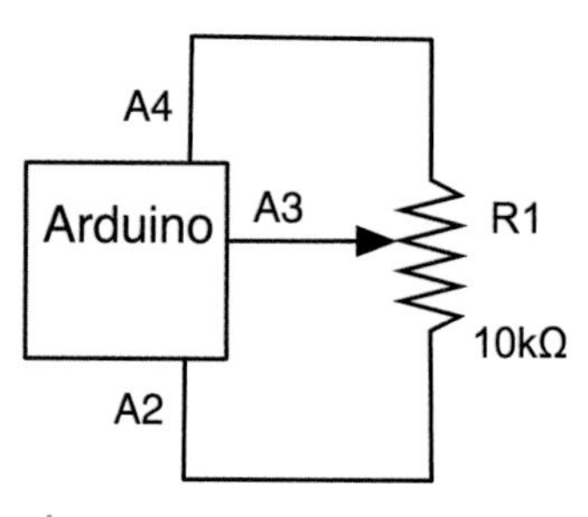

FIGURE 6-17 Schematic diagram—measuring voltage with an Arduino

Software

Load the sketch "voltmeter" into the Arduino IDE and then program your Arduino board with it. Open the Serial Monitor and you should see something like Figure 6-18.

Try twiddling the knob from one end of the range to the other. You should find that you can set the voltage to anything between 0 and 5V.

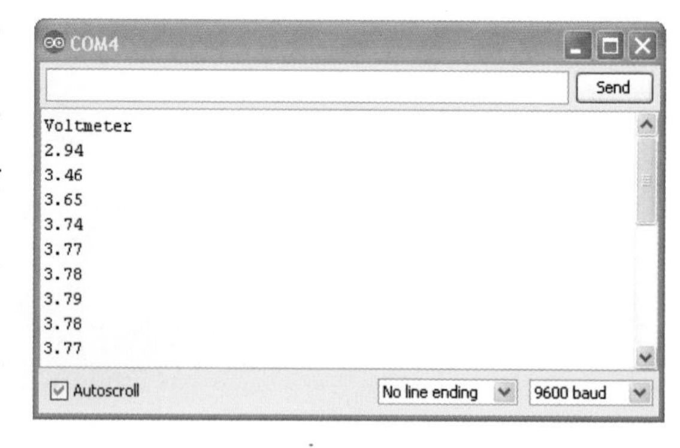

FIGURE 6-18 The Serial Monitor showing voltage at A3

```
// voltmeter

int voltsInPin = 3;
int gndPin = A2;
int plusPin = A4;
```

```
void setup()
{
  pinMode(gndPin, OUTPUT);
  digitalWrite(gndPin, LOW);
  pinMode(plusPin, OUTPUT);
  digitalWrite(plusPin, HIGH);
  Serial.begin(9600);
  Serial.println("Voltmeter");
}

void loop()
{
  int rawReading = analogRead(voltsInPin);
  float volts = rawReading / 204.8;
  Serial.println(volts);
  delay(200);
}
```

The sketch defines the pins as usual. Note that when referring to the analog input pins for use as analog inputs (as with "voltsInPin"), we just use the pin number. So for A3 we specify just 3. However, because we are using A2 and A4 as digital outputs, we have to add the letter A to the front.

The "setup" function sets the pin modes, but also sets "gndPin" and "plusPin" to LOW and HIGH, respectively, before starting serial communication and sending a welcome message.

Inside the loop, we use "analogRead" to give us a raw value of between 0 and 1023, where 0 means 0V and 1023 means 5V. To convert this into an actual voltage, we need to divide it by 204.8 (1023/5). In dividing a raw reading that is an integer (whole number) by a decimal number of 204.8 (called floats in Arduino), the result will be a float, and so we specify the type of the "volts" variable to be "float".

Finally, we print out the voltage and then wait 200 milliseconds before we take the next reading. We don't have to wait before taking the next reading, it just stops the readings from flying up the screen too fast to read.

In the next section, we will use the same hardware with the addition of an external LED and a slightly different sketch to change the rate at which an external LED flashes.

How to Use an Arduino to Control an LED

There are three useful things to be learned here. The first is how to make an Arduino drive an LED. The second is how to control the rate of flashing using a reading from a variable resistor, and finally we will show how to use the Arduino to control the power going to the LED and thus determine its brightness (Figure 6-19).

FIGURE 6-19 An Arduino, variable resistor, and LED

You Will Need

To try this example, you will need the following items.

Quantity	Names	Item	Appendix Code
1		Arduino Uno/Leonardo	M2/M21
1		USB lead; Type B for Uno, Micro USB for Leonardo	
1	R1	10kΩ variable resistor	K1, R1
1	R2	220Ω Resistor	K2
1	D1	LED	K1

Construction

As we discussed in Chapter 4, LEDs need a resistor to stop them drawing too much current. This means we cannot just push one straight into an output pin of an Arduino. So, we are going to start by taking an LED and resistor, shortening the leads, soldering them together and making an LED resistor combo that we can just plug into our Arduino. Figure 6-20 shows the steps involved in making this.

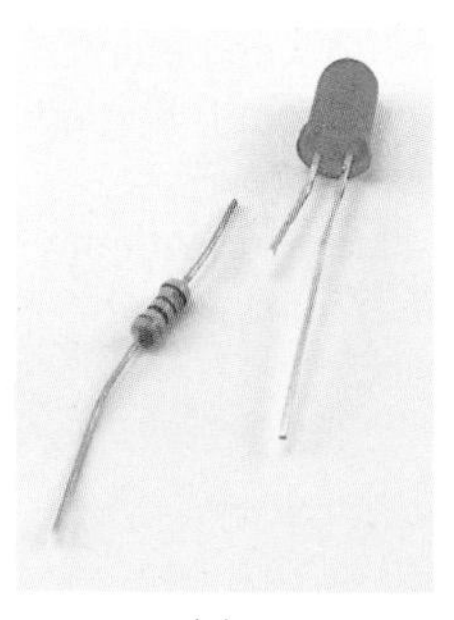

(a)

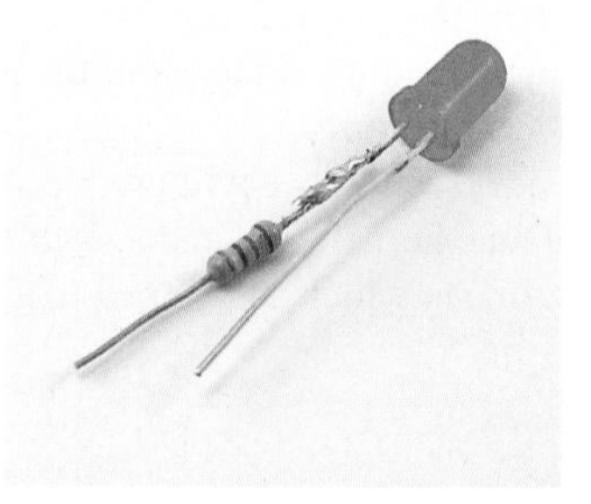

(b)

FIGURE 6-20 Making an LED resistor combo

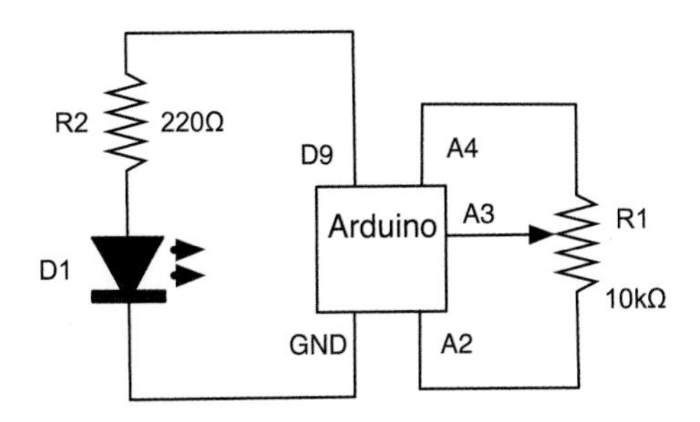

FIGURE 6-21 Schematic for an LED, Arduino, and a variable resistor

To avoid confusion, put the resistor in the anode (positive and longer) lead and keep the combined lead the longer lead, so you know it is the positive end of the combo.

The schematic diagram for the arrangement is shown in Figure 6-21.

We will use pin 9 as a digital output for the LED. The other end of the LED combo being connected to a convenient GND connection.

Keep this Arduino-friendly LED combo. You will use it again in several later sections.

Software (Flashing)

You will use two different sketches with this arrangement of hardware. The first uses the variable resistor to control the speed of the flashing, while the second will control the brightness of the LED.

Attach the LED resistor combo, as shown in Figure 6-19, and load the sketch "variable_led_flash" onto your Arduino board. You should find that turning the knob controls the rate at which the LED flashes.

```
// variable_led_flash

int voltsInPin = 3;
int gndPin = A2;
int plusPin = A4;
int ledPin = 9;

void setup()
{
  pinMode(gndPin, OUTPUT);
  digitalWrite(gndPin, LOW);
  pinMode(plusPin, OUTPUT);
  digitalWrite(plusPin, HIGH);
  pinMode(ledPin, OUTPUT);
}

void loop()
{
  int rawReading = analogRead(voltsInPin);
  int period = map(rawReading, 0, 1023, 100, 500);
  digitalWrite(ledPin, HIGH);
  delay(period);
  digitalWrite(ledPin, LOW);
  delay(period);
}
```

The sketch is quite similar to that in the previous section; however, we no longer use the Serial Monitor, so all that code is gone. We do need to define a new pin "ledPin" to use for the LED.

The "loop" function still reads the raw value from the analog pin A3, but it then uses the "map" function to convert the "rawReading" value of between 0 to 1023 to a range of 100 to 500.

The "map" function is a standard Arduino command that adjusts the range of the value passed in as the first parameter. The second and third parameters are the range of the raw value, while the fourth and fifth are the desired range you want to compress or expand the value into.

We then flash the LED using this number (100 to 500) as the delay between turning the LED on and off. The end result of this is that the LED will flash faster the closer A3 is to 0V.

Software (Brightness)

We can use exactly the same hardware, but with different software to control the brightness of the LED instead of its rate of flashing. This will use the Arduino "analogWrite" function to vary the power going to the pin. This feature is only available for those pins marked with a "~" on the Arduino board. Fortunately, we thought ahead and chose such a pin to connect the LED to.

These pins can use a technique called pulse-width modulation (PWM) to control how much power goes to the output. This works by sending out a series of pulses, around 500 times per second. These pulses may be high for only a short time, in which case little power is delivered, or high until it's nearly time for the next pulse, in which case lots of power is delivered.

In the case of the LED, this means that in each cycle, the LED is either off, on for some of the time, or on the whole time. Our eyes cannot keep up with such a fast-changing event, so it just appears that the brightness of the LED varies.

Load the sketch "variable_led_brightness" onto your Arduino. You should find that, now, the variable resistor controls the brightness of the LED rather than its rate of flashing.

Most of the sketch is the same as the previous one, the difference lies in the "loop" function.

```
void loop()
{
  int rawReading = analogRead(voltsInPin);
  int brightness = rawReading / 4;
  analogWrite(ledPin, brightness);
}
```

The function "analogWrite" expects a value between 0 and 255, so we can take our raw analog reading of between 0 and 1023 and divide it by 4 to put it into roughly the right range.

How to Play a Sound with an Arduino

The first Arduino sketch we tried at the start of this chapter flashed an LED on and off. If we turn a digital output pin on and off much faster than this, we can drive a sounder to create a sound. Figure 6-22 shows a simple sound generator that plays one of two notes when a button is pressed.

You Will Need

To have your Arduino make sounds, you will need the following items.

Quantity	Names	Item	Appendix Code
1		Arduino Uno/Leonardo	M2/M21
1		USB lead; Type B for Uno, Micro USB for Leonardo	
2	S1, S2	Miniature push switches	K1
1	Sounder	Small piezo sounder	M3
1		Breadboard	T5
		Jumper wires or solid-core wire	T6

FIGURE 6-22 A simple Arduino tone generator

Construction

Figure 6-23 shows the schematic for the tone generator, while Figure 6-24 displays the breadboard layout.

Make sure the push switches are the right way around. They should be positioned so the pins extend out of the sides rather than the top and bottom. The piezo sounder may have one pin indicated as positive. Position this at the top of the breadboard.

Attach the components as shown and connect the jumper leads to the Arduino.

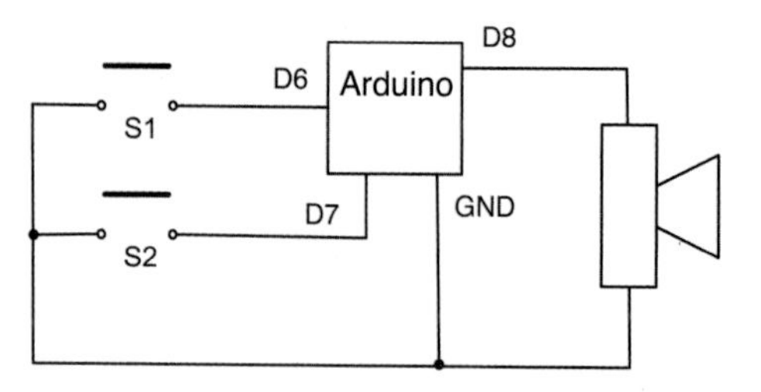

Figure 6-23 Schematic diagram for the tone generator

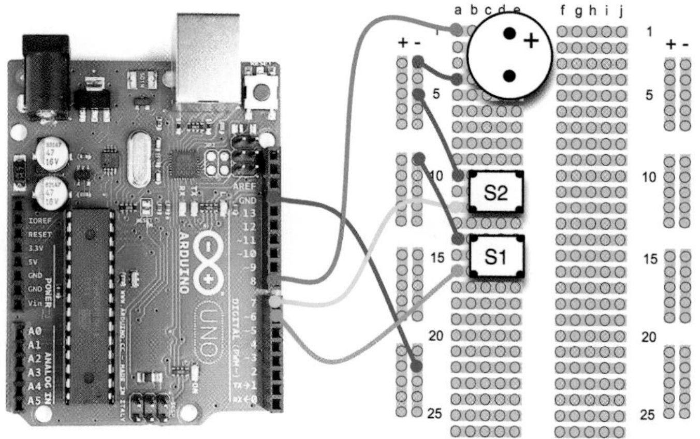

Figure 6-24 Breadboard layout for the tone generator

Software

The sketch is quite straightforward and follows what should be a familiar pattern by now.

```
// arduino_sounds

int sw1pin = 6;
int sw2pin = 7;
int soundPin = 8;

void setup()
{
  pinMode(sw1pin, INPUT_PULLUP);
  pinMode(sw2pin, INPUT_PULLUP);
  pinMode(soundPin, OUTPUT);
}

void loop()
{
  if (! digitalRead(sw1pin))
  {
    tone(soundPin, 220);
  }
  else if (! digitalRead(sw2pin))
  {
    tone(soundPin, 300);
  }
  else
```

```
    {
      noTone(soundPin);
    }
}
```

First, we define the variables for the pins. The switches will be connected to "sw1pin" and "sw2pin". These will be digital inputs, while the "soundPin" will be a digital output.

Note that in the setup function for the switch pins, we use the command "pinMode" with the parameter INPUT_PULLUP. This sets the pin to be an input, but also enables a "pull-up" resistor built into the Arduino, which keeps the input pin HIGH, unless we pull it LOW by pressing the button.

It is because the input pins are normally high that in the "loop" function, when we are checking to see if a button is pressed, we have to use the "!" (logical not). In other words, the following will only sound the tone if the digital input pin "sw1pin" is LOW.

```
if (! digitalRead(sw1pin))
{
tone(soundPin, 220);
}
```

The "tone" function is a useful built-in Arduino function that plays a tone on a particular pin. The second parameter is the frequency of the tone in Hertz (cycles per second).

If no key is pressed, then the function "noTone" is called and stops any tone that is playing.

How to Use Arduino Shields

The success of Arduino had been in no small part due to the wide range of plug-in shields that add useful features to a basic Arduino board. A shield is designed to fit into the header sockets of the main Arduino board. Most shields will then pass through these connections in another row of header sockets, making it possible to construct stacks of shields with an Arduino at the bottom. Shields that have, say, a display on them, will not normally pass through in this way. You also need to be aware that if you stack shields like this, you need to make sure there are no incompatibilities, such as two of the shields using the same pin. Some shields get around this problem by providing jumpers to add some flexibility to pin assignments.

Shield	Description	URL
Motor	Ardumoto shield. Dual H-bridge bidirectional motor control at up to 2A per channel.	www.sparkfun.com/products/9815
Ethernet	Ethernet and SD card shield.	http://arduino.cc/en/Main/ArduinoEthernetShield
Relay	Controls four relays. Screw terminals for relay contacts.	www.robotshop.com/seeedstudio-arduino-relay-shield.html
LCD	16 × 2 character alphanumeric LCD shield with joystick.	www.freetronics.com/products/lcd-keypad-shield

TABLE 6-1 Some Commonly Used Shields

The web site http://shieldlist.org/ will tell you which pins are used by any particular shield.

There are shields available for almost anything you could want an Arduino to do. They range from relay control to LED displays and audio file players.

Most of these are designed with the Arduino Uno in mind, but are also usually compatible with the bigger Arduino Mega and the new Arduino Leonardo.

For an encyclopedic list, that includes useful technical details about the pin usage of these shields, it can be found at http://shieldlist.org/.

Some of the author's favorite shields are listed in Table 6-1.

How to Control a Relay from a Web Page

By using an Ethernet Shield and connecting it to your home hub, you can turn your Arduino into a tiny web server. Since it is still an Arduino, you can still attach electronics to it. So, by using the hacked toy we made in the section "How to Hack a Toy for Arduino Control" and a web interface on the Arduino, we can control our toy over our local network, or if we open up our firewall from the Internet!

Figure 6-25 shows the toy attached to the shield and Arduino along with the browser interface that we will use to control it—first on our computer (Figure 6-25b) and then from a smartphone (Figure 6-25c).

(a)

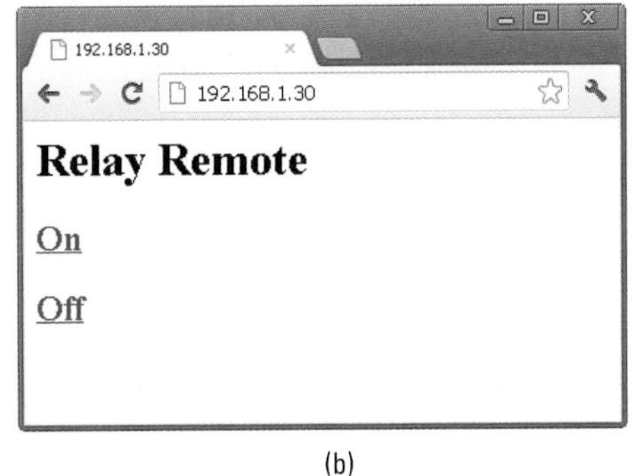

(b)

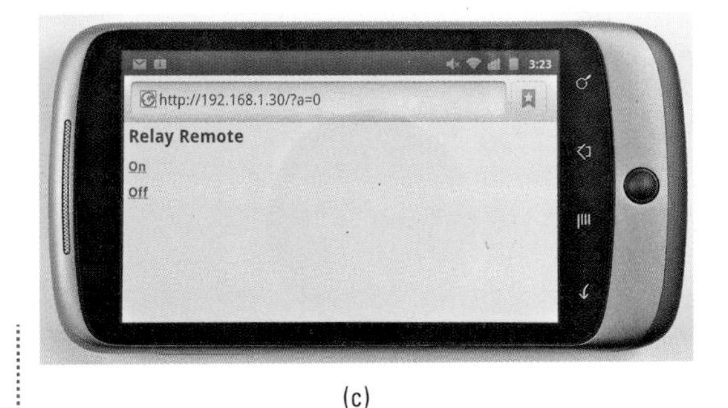

(c)

Figure 6-25 A hacked toy controllable over the network

You Will Need

To make your web-controlled toy, you will first need to complete the section titled "How to Hack a Toy for Arduino Control." In addition, you will need the following items:

Quantity	Item	Appendix Code
1	Arduino Ethernet Shield	M4
1	Ethernet patch cable	T6
1	9V or 12V 500mA Power Supply	M1

Note that this project will only work with the Arduino Leonardo if the Ethernet Shield is the latest R3 Ethernet Shield. If you have an older Ethernet Shield, you will either need to buy a new Ethernet Shield or use an Arduino Uno.

Construction

In this project, the Arduino is powered from an external power supply rather than the USB connection to the computer. There are two reasons for this. One is that the Ethernet Shield will not operate just from USB power, and the other is that once the Arduino has been programmed—there is no need for it to be connected to your computer, so it may as well be powered from a separate adapter.

The wiring diagram for the project is shown in Figure 6-26.

Connect everything up as shown in Figure 6-26 and load the sketch "web_relay" into the Arduino IDE. Don't upload it onto the Arduino itself just yet. There are some configuration changes we need to make first.

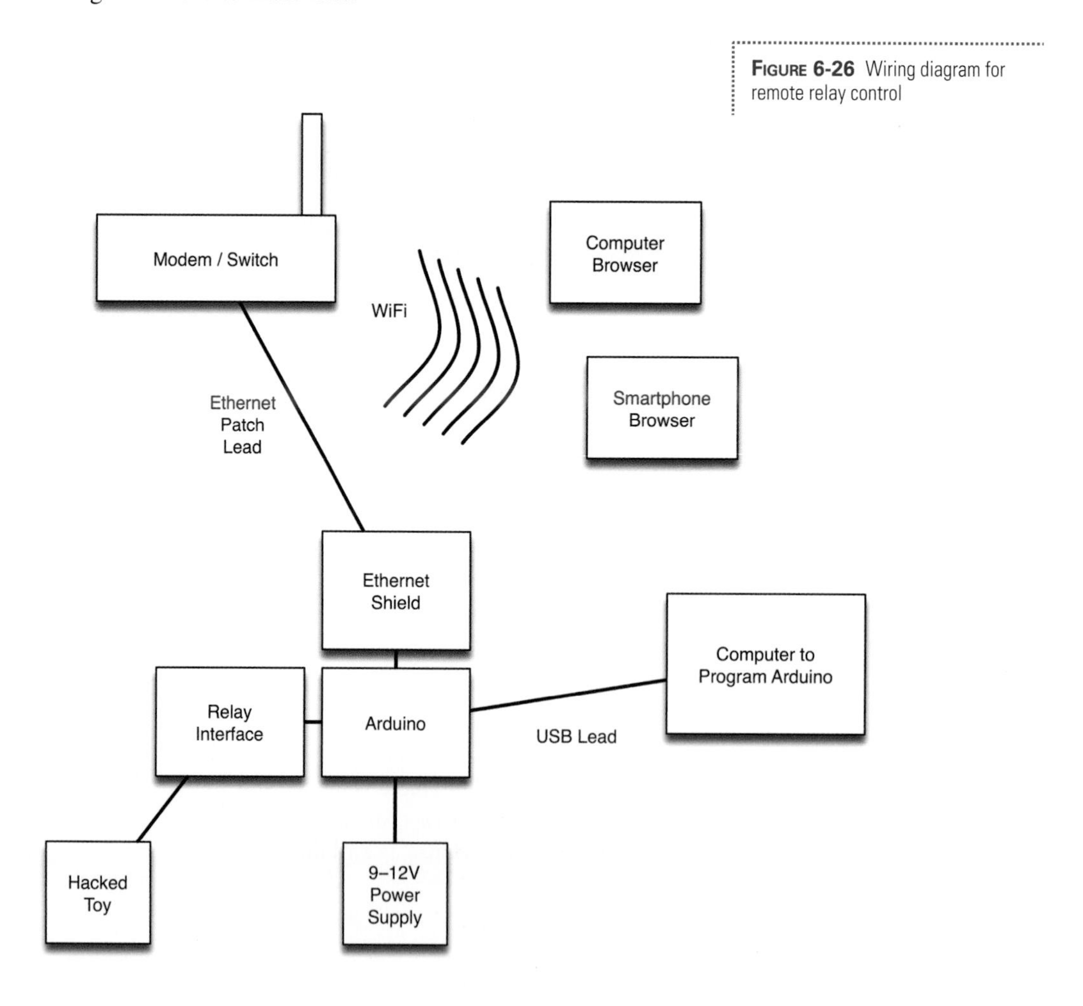

FIGURE 6-26 Wiring diagram for remote relay control

Network Configuration

If you look at the top of the sketch, you will see the lines:

```
byte mac[] = { 0xDE, 0xAD, 0xBE, 0xEF, 0xFE, 0xED };
byte ip[] = { 192, 168, 1, 30 };
```

The first of these, the "mac address" just has to be unique amongst all the devices connected to your network. Note that some of the newer Ethernet Shields have a Mac address printed on them. If you have one of these boards, enter that here. The second one is the IP address. Most devices you connect to your home network will have IP addresses assigned to them automatically by a process called DHCP. That's just fine if you don't really need to know the IP address of your shield and don't mind it changing (say, when using the shield to act like a browser rather than a web server). But in this project, the Arduino and Ethernet Shield are going to act like a web server, so we need to know what the IP address is so we can type it into the address bar of a browser.

You will manually define an IP address. This cannot be any four numbers; they must be numbers that qualify as being internal IP addresses and fit in the range of IP addresses expected by your home router. Typically, the first three numbers will be something like 10.0.1.*x* or 192.168.1.*x*, where *x* is some number between 0 and 255. Some of these IP addresses will be in use by other devices on your network. To find an unused but valid IP address, connect to the administration page for your home router and look for an option that says DHCP. You should somewhere find a list of devices and their IP addresses similar to that in Figure 6-27. Select a final number to use in your IP address. In this case, 192.168.1.30 looked like a good bet and indeed it worked fine.

Set the IP address in the sketch and upload it to your Arduino board.

Testing

Open up a browser on your computer, tablet, or smartphone and navigate to the IP address. If you used the same IP address as the author, that would be http://192.168.1.30. You should then see the web page of Figure 6-25b and 6-25c displayed.

Click the On button and the relay should click on, turning the hacked toy on. The page will then reload in the browser. Clicking "Off" will turn the relay off again.

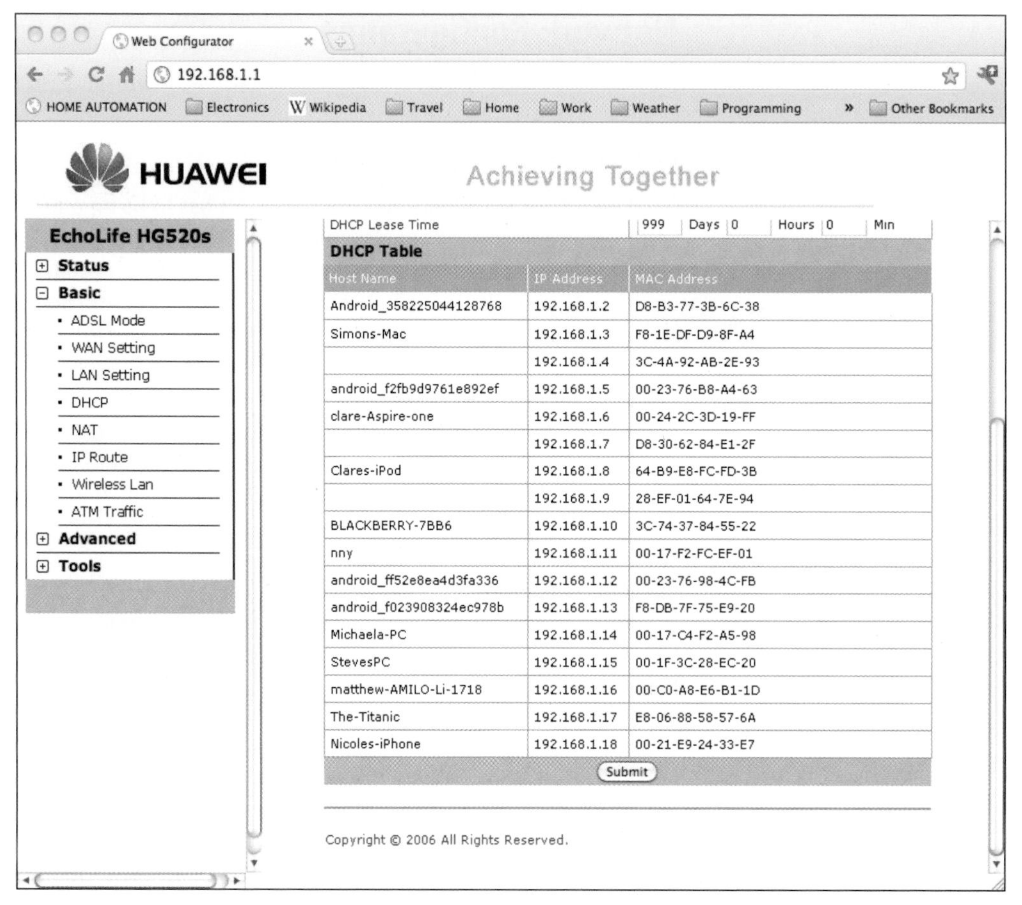

Figure 6-27 Selecting an IP address to use

Software

This sketch is one of the most complicated in the book. However, it is one that can easily be used as a template for other hacks using an Arduino as a web server.

Let's break this sketch down and look at it a section at a time.

```
// web_relay

#include <SPI.h>
#include <Ethernet.h>

// MAC address just has to be unique. This should work
byte mac[] = { 0xDE, 0xAD, 0xBE, 0xEF, 0xFE, 0xED };
// The IP address will be dependent on your local
network:
```

```
byte ip[] = { 192, 168, 1, 30 };
EthernetServer server(80);
int relayPin = A0;
char line1[100];
```

Two libraries have to be included for use with the Ethernet Shield: "SPI" and "Ethernet." Libraries contain a useful collection of functions for, say, a shield. They greatly simplify the writing of sketches as they can just make use of the library functions.

The "SPI" library is for a kind of serial communication that the Arduino uses to send instructions to the Ethernet Shield, and the "Ethernet" library defines some useful functions for us to use with the Ethernet Shield.

After the two variables that define the Mac address and IP address, the next command creates a new "EthernetServer" object that is used every time we want to do something with the Ethernet. We then define the "relayPin" to use and also create a line buffer of 100 characters that is used later in the code when reading the header that arrives from a browser when you navigate to the page being served by the Arduino.

```
void setup()
{
  pinMode(relayPin, OUTPUT);
  Ethernet.begin(mac, ip);
  server.begin();
}
```

The setup function initializes the Ethernet library using the Mac and IP addresses that we set earlier. It also sets the pin mode of the "relayPin" to be an OUTPUT.

```
void loop()
{
  EthernetClient client = server.available();
  if (client)
  {
    while (client.connected())
    {
      readHeader(client);
      if (! pageNameIs("/"))
      {
        client.stop();
        return;
      }
      digitalWrite(relayPin, valueOfParam('a'));
```

```
        client.println("HTTP/1.1 200 OK");
        client.println("Content-Type: text/html");
        client.println();

        // send the body
        client.println("<html><body>");
        client.println("<h1>Relay Remote</h1>");

        client.println("<h2><a href='?a=1'/>On</a></h2>");
        client.println("<h2><a href='?a=0'/>Off</a></h2>");
        client.println("</body></html>");

        client.stop();
      }
    }
  }
```

The loop function is responsible for servicing any requests that come to the web server from a browser. If a request is waiting for a response, then calling "server.available" will return us a "client". If client exists (tested by the first "if" statement), we can then determine if it is connected to the web server by calling "client.connected".

We will come to the "readHeader" function later. This function and "pageNameIs" are used to determine that the browser is actually contacting the page for setting the relay. This is because browsers will often send two requests to a server page, one to try and find an icon for the web site, and a second to the page itself. This code allows us to ignore the icon request.

The next line sets the relay pin using "digitalWrite". The value it sets the output to is whatever value the request parameter "a" is set to. This will be either "1" or "0".

The next three lines of code print out a return header. This just tells the browser what type of content to display. In this case, just HTML.

Once the header has been written, it simply remains to write the remaining HTML back to the browser. This must include the usual "<html>" and "<body>" tags, and also includes a "<h1>" header tag and two "<h2>" tags that are also hyperlinks to this same page, but with the request parameter "a" set to either "0" or "1".

Finally, "client.stop" tells the browser that the message is complete and the browser will display the page.

```
void readHeader(EthernetClient client)
{
  // read first line of header
  char ch;
  int i = 0;
  while (ch != '\n')
  {
    if (client.available())
    {
      ch = client.read();
      line1[i] = ch;
      i ++;
    }
  }
  line1[i] = '\0';
  Serial.println(line1);
}
```

The final three functions in the sketch are general-purpose functions that I tend to use over and over again when making an Arduino web server like this.

The first, "readHeader", reads the header of the request coming from the browser into the buffer "line". We can then use this in the next two functions.

```
boolean pageNameIs(char* name)
{
   // page name starts at char pos 4
   // ends with space
   int i = 4;
   char ch = line1[i];
   while (ch != ' ' && ch != '\n' && ch != '?')
   {
     if (name[i-4] != line1[i])
     {
       return false;
     }
     i++;
     ch = line1[i];
   }
   return true;
}
```

The function "pageNameIs" returns true if the page name part of the header matches the argument supplied. This is what we use in the "loop" function to ignore the icon request from the browser.

```
int valueOfParam(char param)
{
  for (int i = 0; i < strlen(line1); i++)
  {
    if (line1[i] == param && line1[i+1] == '=')
    {
      return (line1[i+2] - '0');
    }
  }
  return 0;
}
```

The "valueOfParam" lets you read the value of the request parameter supplied as an argument. This is much more restricted than the kind of request parameter you will be used to if you have done any web programming. First, the request parameter name must be a single character, and second, its value must be a single digit between 0 and 9. The function will return the value or 0 if there is no parameter of that name.

This is one of those projects that can be adapted for all sorts of purposes.

How to Use an Alphanumeric LCD Shield with Arduino

Another commonly used Arduino shield is the LCD shield (Figure 6-28).

FIGURE 6-28 An LCD shield

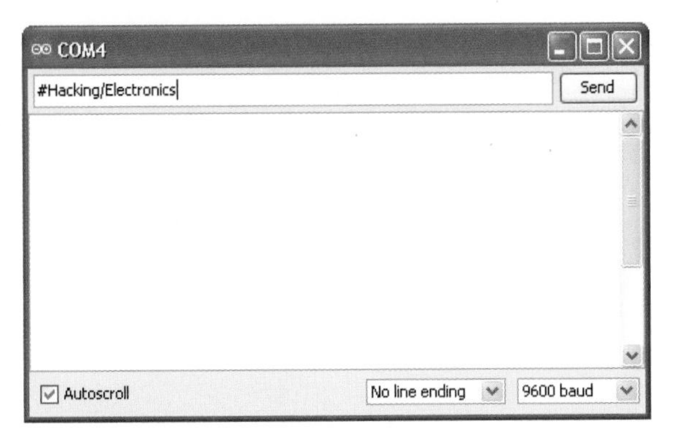

FIGURE 6-29 Sending a message with the Serial Monitor

There are many different shields available and most use an LCD module based on the HD44780 LCD driver chip. The model used here is the Freetronics LCD and Keypad Shield (www.freetronics.com). Most other LCD projects will work with this example code, but you may have to change the pin allocations (discussed later).

This project lets you send a short message (the display is only two lines of 16 characters) using the Serial Monitor (Figure 6-29).

You Will Need

To experiment with an LCD display, you will need the following items.

Quantity	Item	Appendix Code
1	Arduino Uno	M2
1	USB Type A to Type B (as commonly used for USB printers)	
1	LCD Shield	M18

Construction

There is really not very much to construct here. Just plug the LCD shield onto the Arduino and plug in the Arduino to your computer via a USB port.

Software

The software is pretty straightforward too. Again, most of the work is done in the library.

```
// LCD_messageboard

#include <LiquidCrystal.h>

// LiquidCrystal display with:
// rs on pin 8
// rw on pin 11
```

```
// enable on pin 9
// d4-7 on pins 4-7
LiquidCrystal lcd(8, 11, 9, 4, 5, 6, 7);

void setup()
{
  Serial.begin(9600);
  lcd.begin(2, 16);
  lcd.clear();
  lcd.setCursor(0,0);
  lcd.print("Hacking");
  lcd.setCursor(0,1);
  lcd.print("Electronics");
}

void loop()
{
  if (Serial.available())
  {
    char ch = Serial.read();
    if (ch == '#')
    {
      lcd.clear();
    }
    else if (ch == '/')
    {
      lcd.setCursor(0,1);
    }
    else
    {
      lcd.write(ch);
    }
  }
}
```

If you are using a different LCD shield, then check the specification to see which pins it uses for what. You may need to modify the line:

```
LiquidCrystal lcd(8, 11, 9, 4, 5, 6, 7);
```

The parameters to this are the pins that the shield uses for (rs, rw, e, d4, d5, d6, d7). Note that not all shields use the rw pin. If this is the case, just pick the number of a pin not being used for anything else.

The loop reads any input, and if it is a # character, it clears the display. If it is a "/" character, it moves to the second row; otherwise, it just displays the character that was sent.

For example, to send the text displayed in Figure 6-28, you would enter the following into the Serial Monitor:

```
#Hacking/Electronics
```

Notice that the LCD library provides you with the "lcd .setCursor" function to set the position for the next text to be written. The text is then written using the "lcd.write" function.

How to Drive a Servo Motor with an Arduino

Servo motors are a combination of motor, gearbox, and sensor that are often found in remote-controlled vehicles to control steering or the angles of surfaces on remote-controlled airplanes and helicopters.

Unless they are special-purpose servo motors, servo motors do not rotate continuously. They usually only rotate through about 180 degrees, but can be accurately set to any position by sending a stream of pulses.

Figure 6-30 displays a servo motor and shows how the length of the pulses determines the position of the servo.

A servo will have three connections: GND, a positive power supply (5 to 6V), and a control connection. The GND connection is usually connected to a brown or black lead, the

FIGURE 6-30 Controlling a servo motor with pulses

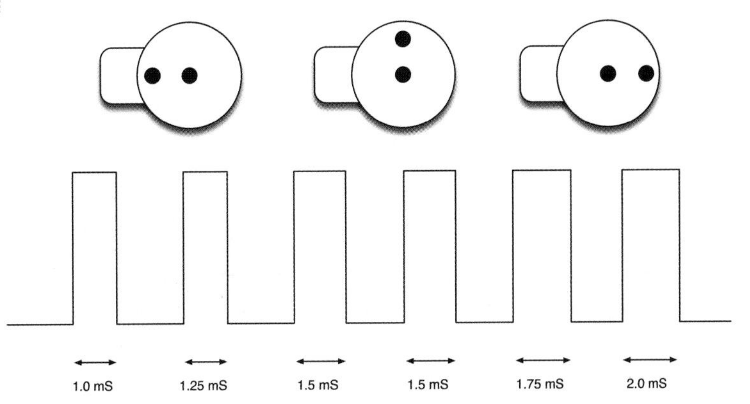

positive connection to a red lead, and the control connection to an orange or yellow lead.

The control connection draws very little current. The server expects to receive a pulse every 20 ms or so. If the pulse is 1.5 ms in duration, then the servo will sit at its middle position. If the pulse is shorter, it will settle in a position to one side, and if the pulse is longer, it will move to a position on the other side of the center position.

You Will Need

To experiment with a servo and Arduino, you will need the following items.

Quantity	Item	Appendix Code
1	Arduino Uno/Leonardo	M2/M21
1	USB lead; Type B for Uno, Micro USB for Leonardo	
1	9g servo motor	H10
1	10kΩ variable resistor	K1, R1
	Jumper wires or solid-core wire	T6

Construction

Figure 6-31 shows a servo connected to an Arduino using a jumper.

Before you power the servo motor from the 5V supply of an Arduino, first check that the Arduino can supply the current requirement. Most small servos will be just fine, such as the tiny 9g servo shown in Figure 6-31.

In Figure 6-31, you can also see the little blue trimpot used to set the position of the servo. This is connected to A1, but uses A0 and A2 to provide GND and +5V to the track ends of the variable resistor.

FIGURE 6-31 Connecting a servo to an Arduino

Software

The Arduino has a library specifically designed for generating the pulses that the servo needs. The following example sketch (called "servo") will use this library to set the position of the servo arm to follow the position of the knob on a variable resistor.

```
// servo

#include <Servo.h>

int gndPin = A0;
int plusPin = A2;
int potPin = 1;
int servoControlPin = 2;
```

After defining the pins to be used, the servo library requires the following line of code to set up the servo.

```
Servo servo;
```

The "setup" function sets up the pins and associates the "servo" with the "servoControlPin".

```
void setup()
{
  pinMode(gndPin, OUTPUT);
  digitalWrite(gndPin, LOW);
  pinMode(plusPin, OUTPUT);
  digitalWrite(plusPin, HIGH);
  servo.attach(servoControlPin);
}
```

The "loop" function continuously reads A1 to determine the position of the variable resistor (a number between 0 and 1023) and divides this number by 6 to convert it to an angle between 0 and 170. This is the angle in degrees to which the servo is then set.

```
void loop()
{
  int potPosition = analogRead(potPin);  // 0 to 1023
  int angle = potPosition / 6;           // 0 to 170
  servo.write(angle);
}
```

How to Charlieplex LEDs

An Arduino only has so many IO pins, so when looking to minimize the number of pins used to display a matrix of LEDs, an interesting technique called Charlieplexing can be used. The name comes from the inventor Charlie Allen at the company Maxim, and the technique takes advantage of the feature of Arduino and other microcontroller IO pins that allows them to be changed from outputs to inputs while a sketch is running.

Figure 6-32 shows the arrangement for controlling six LEDs with three pins.

Table 6-2 shows how the pins should be set to light a particular LED.

LED	Pin 1	Pin 2	Pin 3
A	High	Low	Input
B	Low	High	Input
C	Input	High	Low
D	Input	Low	High
E	High	Input	Low
F	Low	Input	High

TABLE 6-2 Charlieplexing

The number of LEDs that can be controlled per microcontroller pin is given by the following formula:

$$\text{LEDs} = n^2 - n$$

So, if we use four pins, we can have 16 – 4 or 12 LEDs, and 10 pins would give us a massive 90 LEDs, but an awful lot of wiring to do.

There are, however, problems with scaling Charlieplexing up, not the least of which is that the refresh rate needs to be fast enough to fool the eye, and a large number of pins will need a lot of sequence steps to energize all the LEDs that need energizing in a refresh cycle. This will also result in the LEDs becoming dim because their duty cycle will be low. You can compensate for this to some extent by increasing the current through the LEDs, which will cope with fairly large peak currents for a small duration. This does lead to the problem that

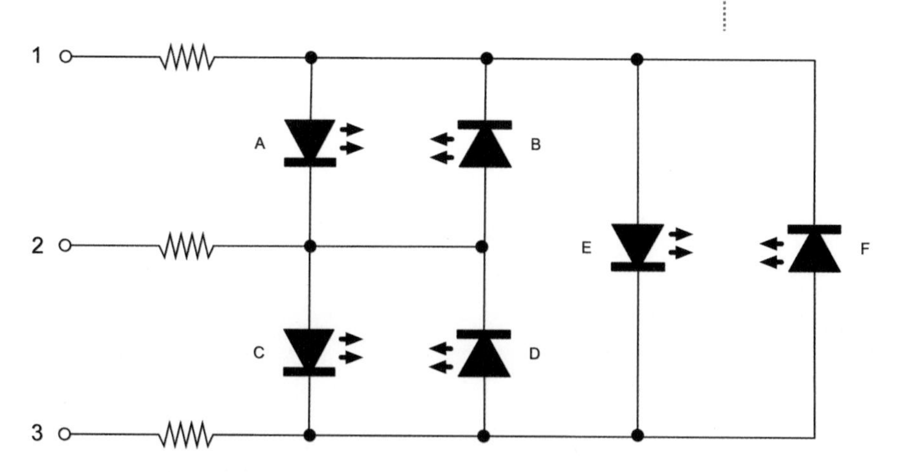

FIGURE 6-32 Charlieplexing LEDs

if the microcontroller freezes for some reason, then LEDs could burn out.

You Will Need

To Charlieplex six LEDs, you will need the following items.

Quantity	Item	Appendix Code
1	Arduino Uno/Leonardo	M2/M21
1	USB lead; Type B for Uno, Micro USB for Leonardo	
6	LEDs	S11
3	220Ω resistors	K2
	Jumper wires or solid-core wire	T6

Construction

You will use breadboard to Charlieplex these LEDs (Figure 6-33).

The breadboard layout for this is shown in Figure 6-34. When constructing it, take special care to get the polarity of each LED correct.

FIGURE 6-33 Six Charlieplexed LEDs on breadboard

The resistors are used to connect Arduino pins D12, D11, and D10 to the breadboard. The resistor leads will stay in the Arduino sockets better if you put a little zigzag kink in the leads using pliers.

The LEDs are very close together, so it is easier to make this with 3mm LEDs. The positive leads of the LEDs (anodes) are shown in red in Figure 6-34.

Software

Load the sketch for this ("charlieplexing") onto your Arduino board, and you should see it cycle through the LEDs in the order A to F, as per Figure 6-32.

The sketch first defines three pins to be used as the control pins.

```
// charlieplexing

int pin1 = 12;
int pin2 = 11;
int pin3 = 10;
```

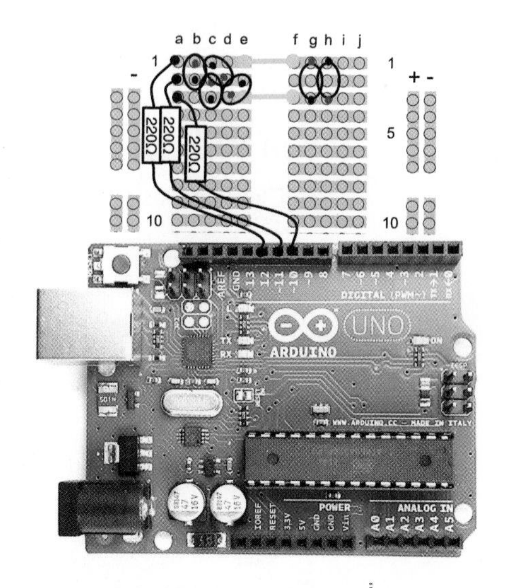

FIGURE 6-34 Breadboard layout for Charlieplexing

The pin states that we need to control each LED, as shown in Table 6-2, are contained in an array called "pinStates". Each element of this array is itself an array of three elements, one for each control pin. A value of 1 indicates that the control pin should be HIGH for that LED, 0 for a LOW, and –1 for an INPUT.

```
int pinStates[][3] = {
  {1, 0, -1}, // A
  {0, 1, -1}, // B
  {-1, 1, 0}, // C
  {-1, 0, 1}, // D
  {1, -1, 0}, // E
  {0, -1, 1}  // F
};
```

As we will be changing the pin mode of the control pins as we go, there is nothing to put in the "setup" function. You do, however, have to have the function there, even if it has nothing in it.

```
void setup()
{
}
```

The loop function steps through each of the LEDs and sets the control pin states according to the values in the appropriate row of the array, using the function "setPins".

```
void loop()
{
  for (int i = 0; i < 6; i++)
  {
    setPins(pinStates[i][0], pinStates[i][1], pinStates[i][2]);
    delay(1000);
  }
}
```

The function "setPins" does not do much except conveniently put the commands to set the state for each control pin into a single convenient line. Most of the logic lives in the "setPin" function that it calls.

```
void setPins(int p1, int p2, int p3)
{
  setPin(pin1, p1);
  setPin(pin2, p2);
  setPin(pin3, p3);
}
```

The "setPin" function sets the state of the pin supplied as its first argument. If the value of the state is –1, then the pin mode for the pin is set to INPUT. Otherwise, it is assumed to be a 1 or a 0 and the pin is set to be an OUTPUT and set to the value supplied with a call to "digitalWrite".

```
void setPin(int pin, int value)
{
  if (value == -1)
  {
    pinMode(pin, INPUT);
  }
  else
  {
    pinMode(pin, OUTPUT);
    digitalWrite(pin, value);
  }
}
```

How to Type Passwords Automatically

The Arduino Leonardo can be used to impersonate a USB keyboard. Unfortunately, this is not true of the Arduino Uno, so in this section you will need an Arduino Leonardo.

Figure 6-35 shows the device we are going to construct.

All that happens when you press the button is that the Arduino Leonardo pretends to be a keyboard and types the password set in the sketch, wherever the cursor happens to be.

Figure 6-35 Entering passwords automatically with Arduino Leonardo

You Will Need

To build this, you will need the following items.

Quantity	Item	Appendix Code
1	Arduino Leonardo	M21
1	Micro USB lead for the Leonardo	
1	Impressive switch	H15
	Hookup wire	T7

Construction

Solder leads to the switch, and tin the ends so they can be pushed directly into the sockets on the Arduino. One lead from the switch should go to digital pin 2 and the other to GND.

Program the Arduino Leonardo with the sketch "password". Note that when programming the Leonardo, you may have to hold down the reset button until the message "uploading…" appears in the Arduino software.

Software

To use the project, just position your mouse over a password field and press the button. Please note that this project is really just to illustrate what you can do with an Arduino Leonardo. To find your password, all someone would have to do is press the button while in a word processor. So, in terms of security, it is about as secure as writing your password on a sticky note and attaching it to your computer monitor!

The sketch is very simple. The first step is to define a variable to contain your password. You will need to change this to your password. We then define the pin to use for the switch.

```
// password
// Arduino Leonardo Only

char* password = "mysecretpassword";

const int buttonPin = 2;
```

The Leonardo has access to special keyboard and mouse features not available to other types of Arduino. So, in the "setup" function, the Keyboard feature is started with the line "Keyboard.begin()".

```
void setup()
{
  pinMode(buttonPin, INPUT_PULLUP);
  Keyboard.begin();
}
```

In the main loop, the button is checked with a digital read. If the button is pressed, then the Leonardo uses "Keyboard.print" to send the password. It then waits two seconds to prevent the password being sent multiple times.

```
void loop()
{
  if (! digitalRead(buttonPin))
  {
    Keyboard.print(password);
    delay(2000);
  }
}
```

Summary

This chapter should have got you started using the Arduino and given you some food for thought for clever hacks using it. It has, however, only scratched the surface of what is possible with this versatile board.

For more information on programming the Arduino, you may wish to look at some of the author's other books on this topic. *Programming Arduino: Getting Started with Sketches* assumes no prior programming experience and will show you how to program the Arduino from first principals. *30 Arduino Projects for the Evil Genius* is a project-based book that explains both the hardware and programming side of Arduino, and is illustrated with example projects, nearly all of which are built on breadboard.

The official Arduino web site, www.arduino.cc, has a wealth of information on using the Arduino, as well as the official documentation for the Arduino commands and libraries.

7

Hacking with Modules

There are many modules available that provide a great shortcut when hacking together a project. These modules are usually a tiny PCB with a few components on them and some convenient connection points. They make it very easy to use some surface-mounted ICs that would otherwise be very difficult to solder connections to. Many of these modules are designed to be used with microcontrollers like the Arduino.

In this chapter, you will explore some of the more fun and useful modules available from suppliers like SparkFun and Adafruit, most of whose modules are also open-source hardware. So you'll get to see the schematics for them and even make your own modules using the design if you wish.

Access to the schematics and data sheets is very useful when trying to use a module. There are a few important things you need to know about any module before you use it:

- What is the range of supply voltage?
- How much current does it consume?
- How much current can any outputs supply?

How to Use a PIR Motion Sensor Module

PIR motion sensors are used in intruder alarms and for automatic security alarms. They detect movement using infrared light. They are also cheap and easy to use.

In this example, you will first experiment with a PIR module using it to light an LED, and then look at how it could be hooked up to an Arduino to send a warning message to the Serial Console.

You Will Need (PIR and LED)

Quantity	Names	Item	Appendix Code
1		PIR module (5–9V)	M5
1	D1	LED	K1
1	R1	470Ω resistor	K2
1		Solderless breadboard	T5
		Solid-core jumper wire	T6
1		4 × AA battery holder	H1
1		4 × AA batteries	
1		Battery clip	H2

Breadboard

Figure 7-1 shows the schematic diagram for this experiment.

Looking at the datasheet for this particular module, the supply voltage range is 5V to 7V, so it will work just fine with our four AA batteries.

The module is very easy to use. You just supply it with power and its output goes high (to supply voltage) when movement is detected and then back low again after a second or two.

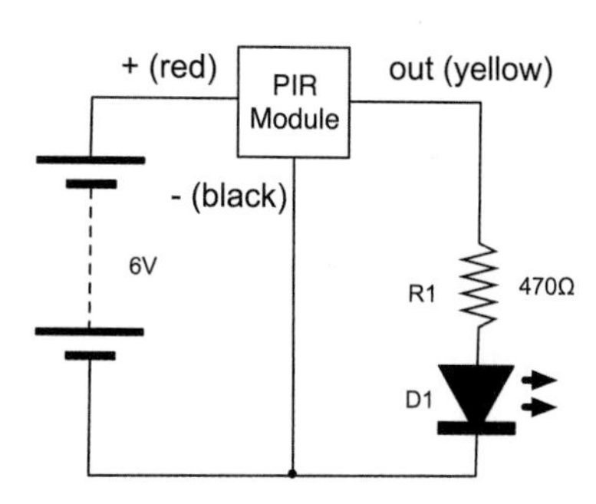

FIGURE 7-1 Schematic diagram—using a PIR module with an LED

The datasheet also says that the output can supply up to 10mA. That isn't a great deal, but is enough to light an LED. By choosing a 470Ω resistor, we will be limiting the current to:

$$I = V / R = (6V - 2V) / 470\Omega = 4 / 470 = 8.5mA$$

Figure 7-2 shows the breadboard layout, while Figure 7-3 offers a photograph of the actual breadboard.

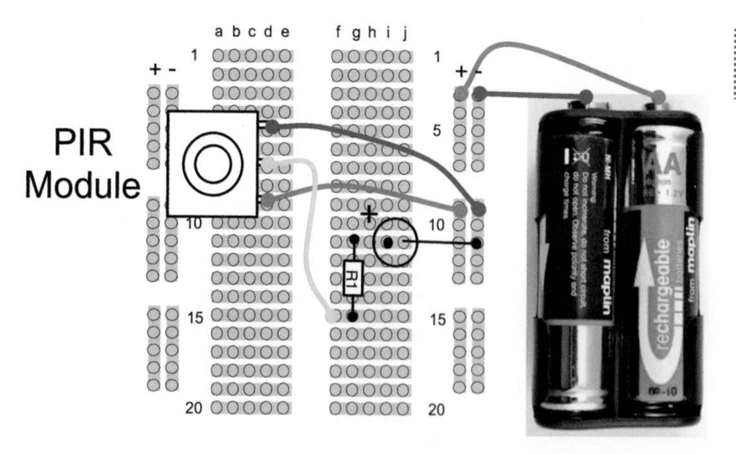

FIGURE 7-2 Breadboard layout—using a PIR module with an LED

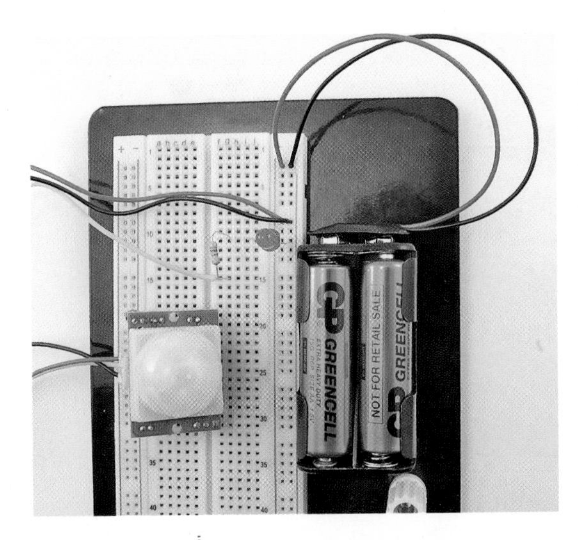

FIGURE 7-3 Using a PIR module with an LED

The PIR module has three pins labeled +5V, GND, and OUT. The supplied connector lead has red, black, and yellow leads. Hook it up so the red lead connects to the connection labeled +5V.

When it's powered up, the LED will light every time movement is detected.

Having already discussed the PIR sensor so we know what to expect of it, it's time to interface it with an Arduino.

You Will Need (PIR and Arduino)

To interface the PIR sensor with an Arduino, you really only need the PIR sensor and an Arduino.

Quantity	Item	Appendix Code
1	PIR module (5–9V)	M5
1	Arduino Uno/Leonardo	M2/M21
1	USB lead; Type B for Uno, Micro USB for Leonardo	

Construction

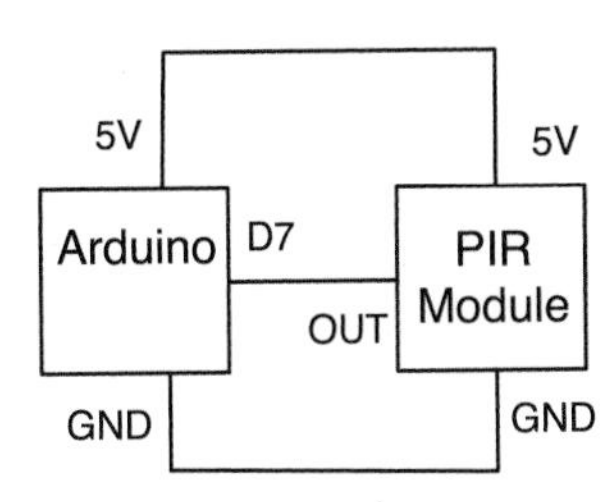

FIGURE 7-4 Schematic diagram for the Arduino and PIR sensor

Figure 7-4 shows the schematic diagram for this, while Figure 7-5 shows how the PIR module is wired to the module. To get the wires to stay in the Arduino sockets, it helps to put a little zigzag bend in the tinned end of the wire lead.

Before you move onto the next stage of programming the Arduino, temporarily remove the OUT lead from its Arduino socket. The reason for this is that you do not know what sketch was last running on the Arduino. It might have been something where pin 7 was an output, and if it was, this could easily damage the output electronics of the PIR sensor.

Figure 7-5 The Arduino and the PIR sensor

Software

Load the sketch "pir_warning" into the Arduino IDE and onto the Arduino board, and then plug the yellow "OUT" lead back into pin 7 on the Arduino.

When you launch the Serial Monitor (Figure 7-6), you will see an event appear every time movement is detected. Imagine leaving this running while away from your computer—to detect snoopers!

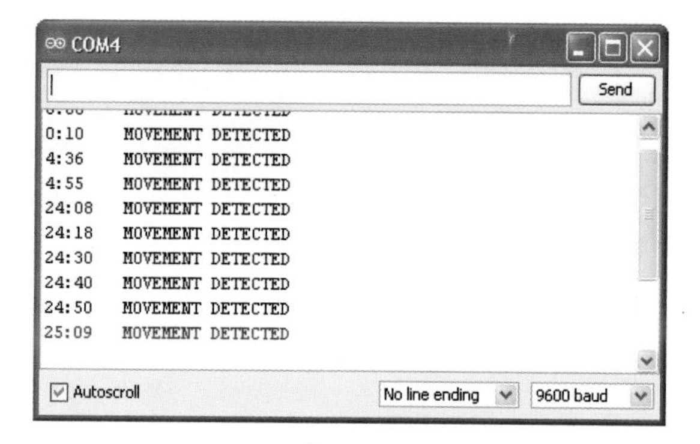

Figure 7-6 The Serial Monitor showing intruder alerts

The sketch is very straightforward.

```
// pir_warning

int pirPin = 7;

void setup()
{
  pinMode(pirPin, INPUT);
  Serial.begin(9600);
}

void loop()
{
  if (digitalRead(pirPin))
  {
    int totalSeconds = millis() / 1000;
```

```
    int seconds = totalSeconds % 60;
    int mins = totalSeconds / 60;
    Serial.print(mins);
    Serial.print(":");
    if (seconds < 10) Serial.print("0");
    Serial.print(seconds);
    Serial.println("\tMOVEMENT DETECTED");
    delay(10000);
  }
}
```

The only part of the code that is a bit different than the other sketches we have seen deals with displaying an elapsed time in minutes and seconds next to each event.

This code uses the Arduino "millis" function, which returns the number of milliseconds since the Arduino was last reset. This is then separated into its minute and second components and the various parts printed out as a message. The last part to be displayed uses the "println" command that adds a line feed to the end of the text so the next text starts on a new line.

The special character "\t" in this "println" is a tab character, to line the output up neatly.

How to Use Ultrasonic Rangefinder Modules

Ultrasonic rangefinders use ultrasound (higher frequency than the human ear can hear) to measure the distance to a sound-reflective object. They measure the time it takes for a pulse of sound to travel to the object and back. Figure 7-7 shows two different types of sonar. On the left is a low-cost sonar module (less than USD 5) with separate ultrasonic transducers for

FIGURE 7-7 Ultrasonic rangefinders

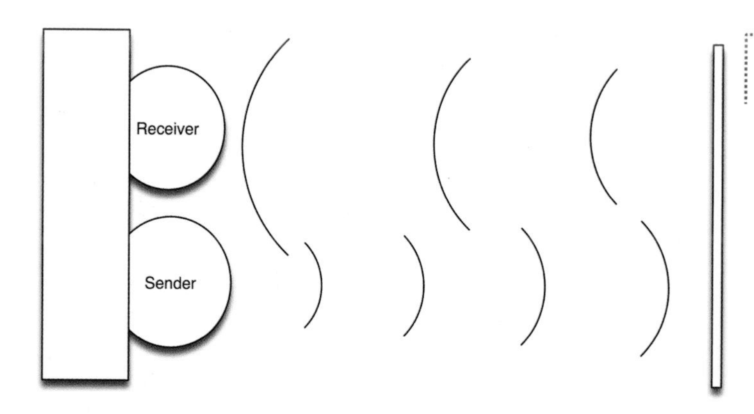

FIGURE 7-8 Ultrasonic range finding

sending the pulse and receiving the echo, and a much more expensive (around USD 25) but highly specified module made by MaxBotix Inc.

To see how to use these modules with an Arduino, we will try out each in turn.

Ultrasonic range finding works the same as sonar used by ships and submarines. A sound wave is sent out from a sender, hits an object, and bounces back. Since we know the speed of sound, the distance to the sound-reflecting object can be calculated from the time it takes for the sound to come back to the receiver (Figure 7-8).

The sound used is at a high frequency—hence, it is called ultrasonic. Most units operate at a frequency of about 40 kHz. Not many people can hear sounds above 20 kHz.

You Will Need

To try out both rangefinders, you will need the following items.

Quantity	Item	Appendix Code
1	Arduino Uno/Leonardo	M2/M21
1	USB Lead; Type B for Uno, Micro USB for Leonardo	
1	MaxBotix LV-EZ1 rangefinder	M6
1	HC-SR04 rangefinder	M7
1	Solderless breadboard	T5
	Solid-core jumper wire	T6

The HC-SR04 Rangefinder

FIGURE 7-9 An HC-SR04 rangefinder on an Arduino

These modules make the Arduino do a lot of the work, which is one reason why they are so much cheaper than the MaxBotix modules. They do, however, have the advantage that they can just fit into the side connector of an Arduino if you can spare the pins to supply them with current using two output pins (Figure 7-9).

Load the sketch "range_finder_budget" onto the Arduino and then plug the rangefinder module into the Arduino, as shown in Figure 7-9.

When you open the Serial Monitor, you will see a stream of distances in inches appear (Figure 7-10). Try pointing the rangefinder in different directions—say, a wall a few feet away—and confirm that the reading is reasonably accurate with a tape measure.

The Arduino code for measuring the range is all contained within the "takeSounding_cm" function. This sends a single 10-microsecond pulse to the "trigger" pin of the ultrasonic module, which then uses the built-in Arduino function "pulseIn" to measure the time period before the echo pin goes high.

FIGURE 7-10 Distance readings in the Serial Monitor

```
// range_finder_budget

int trigPin = 9;
int echoPin = 10;
int gndPin = 11;
int plusPin = 8;

int lastDistance = 0;

void setup()
{
  Serial.begin(9600);
  pinMode(trigPin, OUTPUT);
  pinMode(echoPin, INPUT);
  pinMode(gndPin, OUTPUT);
  digitalWrite(gndPin, LOW);
  pinMode(plusPin, OUTPUT);
  digitalWrite(plusPin, HIGH);
}
```

```
void loop()
{
  Serial.println(takeSounding_in());
  delay(500);
}

int takeSounding_cm()
{
  digitalWrite(trigPin, LOW);
  delayMicroseconds(2);
  digitalWrite(trigPin, HIGH);
  delayMicroseconds(10);
  digitalWrite(trigPin, LOW);
  delayMicroseconds(2);
  int duration = pulseIn(echoPin, HIGH);
  int distance = duration / 29 / 2;
  if (distance > 500)
  {
    return lastDistance;
  }
  else
  {
    lastDistance = distance;
    return distance;
  }
}

int takeSounding_in()
{
  return takeSounding_cm() * 2 / 5;
}
```

We then need to convert that time in milliseconds into a distance in centimeters. If there is no reflection because there is no object that is close enough, or the object is reflecting the sound wave away rather than letting it bounce back to the receiver, then the time of the pulse will be very large and so the distance will also be recorded as very large.

To filter out these long readings, we disregard any measurement that is greater than 5m, returning that last sensible reading we got.

The speed of sound is roughly 343 m/s in dry air at 20 degrees C, or 34,300 cm/s.

Or, put another way, 34,300 / 1,000,000 cm / microsecond.

That is, 0.0343 cm/microsecond.

Put another way, 1/0.0343 microseconds/cm.

Or, 29.15 microseconds/cm.

Thus, a time of 291.5 microseconds would indicate a distance of 10 cm.

The "takeSounding_cm" function approximates 29.15 to 29 and then also divides the answer by 2, as we don't want the distance of the whole return journey, just the distance to the subject.

In actual fact, many factors affect the speed of sound, so this approach will only ever give an approximate answer. The temperature and the humidity of the air will both affect the measurement.

The MaxBotix LV-EZ1 Rangefinder

The HC-SR04 rangefinder only has a single type of interface, and we have to tell it to generate the sonar pulse, and then time how long it takes to come back ourselves.

In contrast, the MaxBotix device does all of this for us, and what's more it provides us with no less than three ways to get the distance readings:

- Serial data readings
- Analog (Vcc / 512) / inch
- Pulse width (147 µS/inch)

FIGURE 7-11 The MaxBotix rangefinder and Arduino

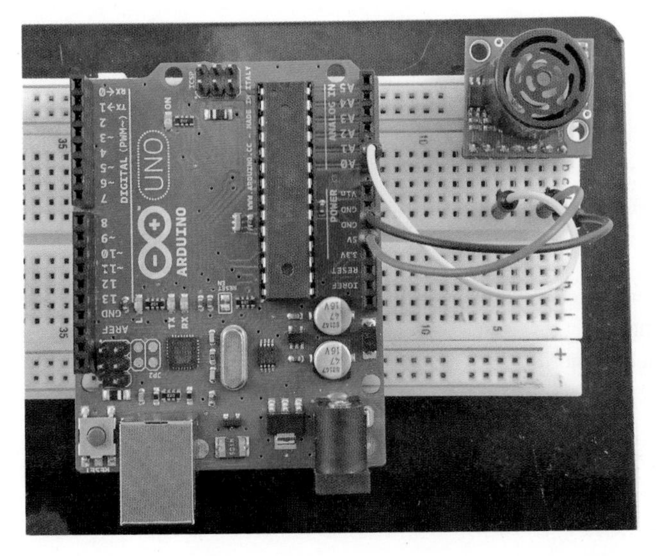

We will use the analog method to test out this device. The figure of Vcc / 512 per inch means that the analog output will be the supply voltage divided by 512 per inch. So if the object was 10 inches away, then the analog output voltage would be:

10 inches × 5V/ 512 = 0.098V

The MaxBotix module has too many pins to easily fit directly onto the Arduino connectors, so you will need to use breadboard.

Figure 7-11 shows the unit and the Arduino on the breadboard, while Figure 7-12 shows the breadboard layout.

Load up the sketch "range_finder_maxsonar" and then connect up the module as shown in Figure 7-11.

The sketch is much simpler than for the other module, and the distance in inches is just the raw analog reading (between 0 and 1023) divided by two.

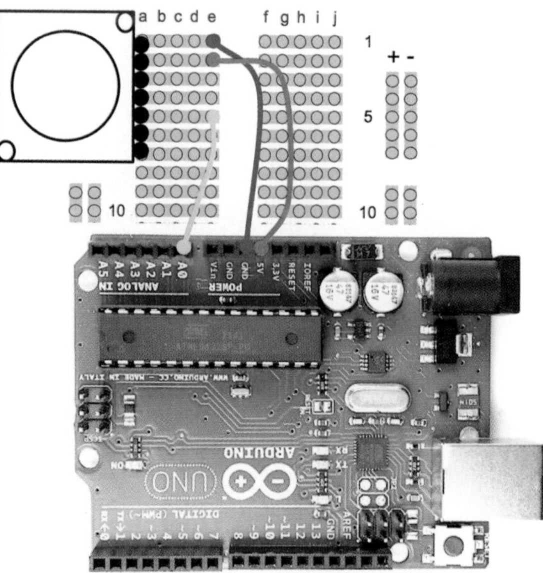

FIGURE 7-12 The MaxBotix rangefinder and Arduino breadboard layout

```
// range_finder_maxsonar

int readingPin = 0;

int lastDistance = 0;

void setup()
{
  Serial.begin(9600);
}

void loop()
{
  Serial.println(takeSounding_in());
  delay(500);
}

int takeSounding_in()
{
  int rawReading = analogRead(readingPin);
  return rawReading / 2;
}

int takeSounding_cm()
{
  return takeSounding_cm() * 5 / 2;
}
```

Opening the Serial Monitor will produce the same stream of distance measurements as the other module.

Note that both sketches have a metric and imperial distance measurement flavor for your convenience.

How to Use a Wireless Remote Module

Radio frequency circuits usually are not worth making yourself when extremely useful modules like the one shown in Figure 7-13 are readily available for just a few dollars.

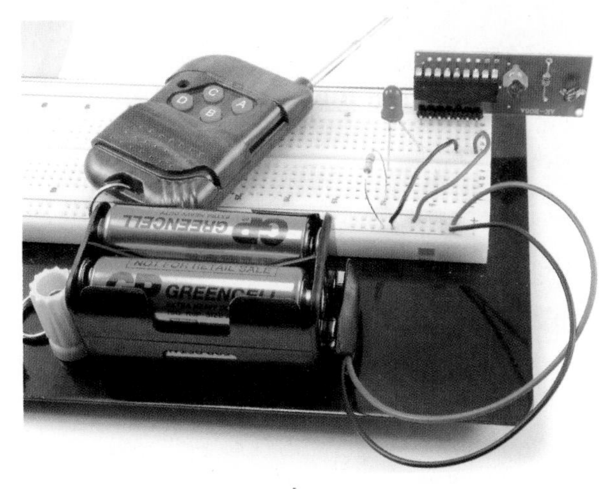

FIGURE 7-13 An RF module on breadboard

The module shown can be easily found on eBay and has a handy little key-fob sized remote with four buttons on it. These buttons can toggle four digital pins on and off on the corresponding receiver module.

It is worth noting that modules like this are also available with relays instead of digital outputs, making it very easy to hack your own remote control projects.

You will first experiment with the module on breadboard, just turning on an LED, and then in the following section, you can try connecting it to an Arduino.

You Will Need

To try out the wireless remote on breadboard, you will need the following items.

Quantity	Names	Item	Appendix Code
1		Solderless breadboard	T5
		Solid-core jumper wire	T6
1		Wireless remote kit	M8
1	D1	LED	K1
1	R1	470Ω resistor	K2
1		4 ×AA battery holder	H1
1		Battery clip	H2
4		AA batteries	

Breadboard

Figure 7-14 shows the breadboard layout used to test the remote. You could, if you wished, add three more LEDs so there was one for each channel.

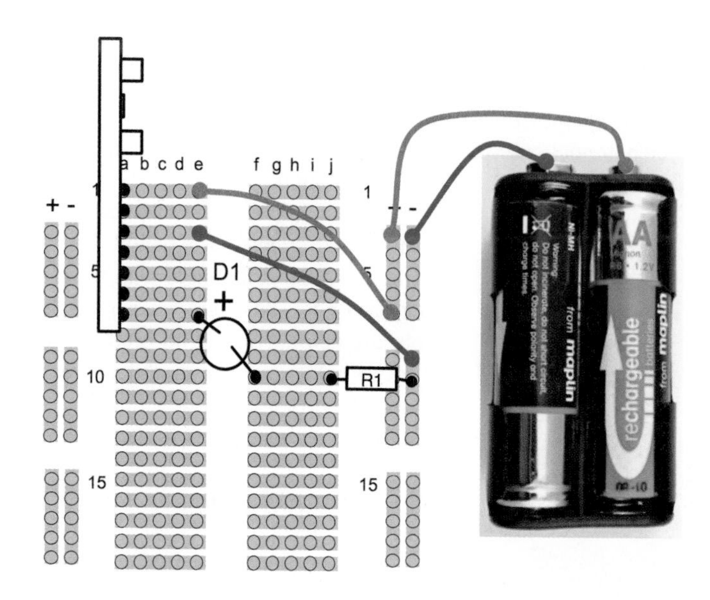

FIGURE 7-14 Breadboard layout for testing the RF module

The datasheet for this module shows that the pins are as shown in Table 7-1.

Put the module on the breadboard with pin 1 at the top of the breadboard, and wire it up as shown in Figure 7-14.

That really is all there is to it. Pressing button A should toggle the LED on and off. If you wanted to, you could add more LEDs so there was one for each channel, or try moving the LED to a different output to check that they all work.

Pin Number	Pin Name	Purpose
1	Vcc	Positive supply 4.5 to 7V
2	VT	Switch voltage—no connection needed
3	GND	Ground
4	D3	Digital output 3
5	D2	Digital output 2
6	D1	Digital output 1
7	D0	Digital output 0

TABLE 7-1 RF Receiver Pinout

How to Use a Wireless Remote Module with Arduino

If we are prepared to lose one of the four channels of the remote from the section "How to Use a Wireless Remote Module," then we can plug the receiver straight into the Arduino socket A0 to A5 (see Figure 7-15).

You Will Need

To try out the wireless remote with an Arduino, you will need the following items.

Quantity	Item	Appendix Code
1	Arduino Uno/Leonardo	M2/M21
1	USB lead; Type B for Uno, Micro USB for Leonardo	
1	Wireless Remote Kit	M8

Before plugging the remote receiver into the Arduino, upload the sketch "rf_remote".

Software

With the software uploaded and the RF receiver attached, when you open the Serial Monitor you should see something like Figure 7-16.

FIGURE 7-15 Using a RF remote with an Arduino

The sketch displays as a 1 or 0 the current state of the remote control channels. So button A will not do anything (that is the button we sacrificed), but pressing the other buttons should toggle the appropriate column between 0 and 1.

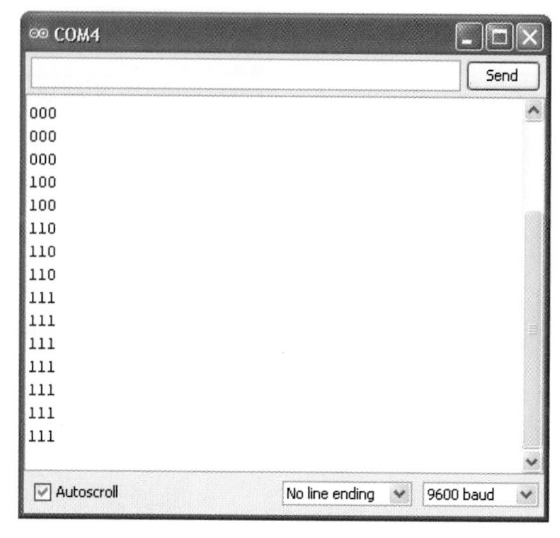

FIGURE 7-16 Remote control messages to your computer

```
// rf_remote

int gndPin = A3;
int plusPin = A5;
int bPin = A2;
int cPin = A1;
int dPin = A0;

void setup()
{
  pinMode(gndPin, OUTPUT);
  digitalWrite(gndPin, LOW);
  pinMode(plusPin, OUTPUT);
  digitalWrite(plusPin, HIGH);
  pinMode(bPin, INPUT);
  pinMode(cPin, INPUT);
  pinMode(dPin, INPUT);
  Serial.begin(9600);
}

void loop()
{
  Serial.print(digitalRead(bPin));
  Serial.print(digitalRead(cPin));
  Serial.println(digitalRead(dPin));
  delay(500);
}
```

The RF receiver uses very little current, so there is no problem powering it from a digital output. In fact, doing so has the added benefit that we can actually turn it off to save power simply by setting the "plusPin" low.

How to Control Motor Speed with a Power MOSFET

This section is a little out of place because MOSFET transistors are not modules. However, this section does lead in nicely to the next section on using motor controller modules.

We first used power MOSFETs in Chapter 3. They are a kind of transistor that is particularly suited to switching high-current loads efficiently. By efficiently, I mean that they run pretty cool, working very well as electronic switches. They have a very low "on" resistance and a very high "off resistance."

Back in Chapter 6, you used a technique called PWM (pulse-width modulation) to control the brightness of an LED by varying the length of pulses. You can use exactly the same trick on a DC motor. However, unlike an LED, motors use too much current to be driven directly from an Arduino output, so you will use a MOSFET controlled by the Arduino.

You Will Need

To build this, you will need the following items.

Quantity	Name	Item	Appendix Code
1		Solderless breadboard	T5
		Solid-core jumper wire	T6
1		4 × AA battery holder	H1
1		4 × AA batteries	
1		Battery clip	H2
1	R1	10kΩ trimpot	K1
1	R2	1kΩ resistor	K2
1	T1	FQP30N06 MOSFET	S6
1		6V DC motor or gear motor	H6
1		Arduino Uno/Leonardo	M2/M21
1		USB lead; Type B for Uno, Micro USB for Leonardo	

The DC motor can be any small motor you can find that is around 6V.

Breadboard

Figure 7-17 shows the schematic diagram.

Notice that we actually have two sources of power here. We have the Arduino, which will be getting its power from your computer's USB port and a separate battery that supplied the power to the MOSFET. This is quite a common arrangement, because the Arduino's 5V output is not really suitable for high-current loads like a motor. Indeed, motors can create all sorts of problems for delicate electronics, so it is best not to power them from the Arduino.

There is less of a problem if the Arduino and motor share a power supply. For example, a 9V battery provides power to the Arduino through its power jack, and at the same time provides the positive supply to the motor.

I have included a resistor R2 between the Arduino output pin and the MOSFET. The circuit would work fine just connecting D5 directly to the gate; however, the gate acts like a capacitor, which means that when switched at very high speed it can actually cause quite a lot of current to flow from the digital output. This will not be a problem at the relatively slow PWM speeds used by the Arduino, but it is considered "good practice" to use a resistor here.

Figures 7-18 and 7-19 show the actual circuit and the breadboard layouts, respectively.

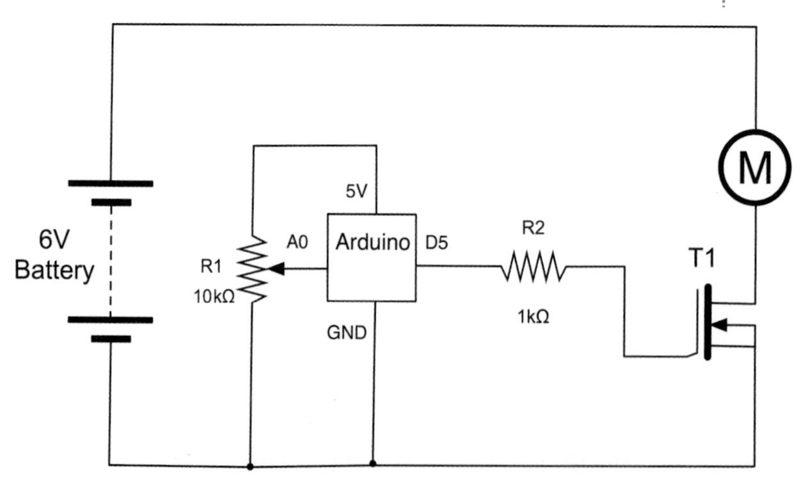

FIGURE 7-17 Schematic diagram for the MOSFET motor control

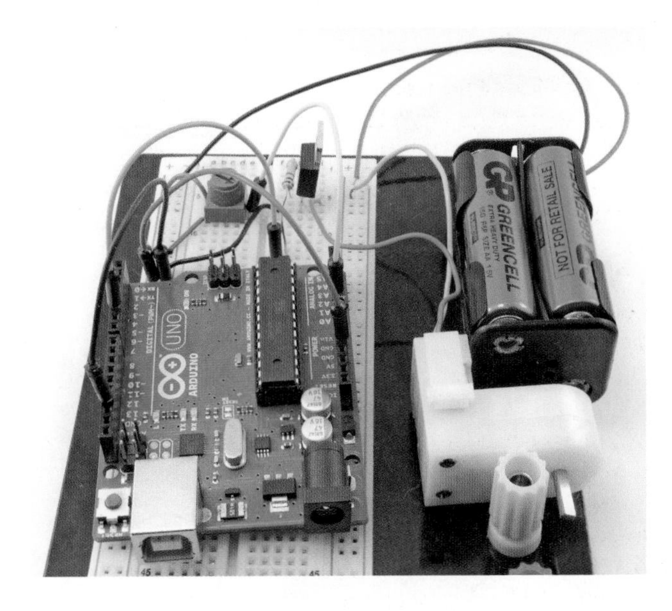

FIGURE 7-18 The MOSFET motor control

Software

Load the sketch "mosfet_motor_speed" onto the Arduino and connect the battery. You should find that by turning the variable resistor you have much finer control of the motor's speed than you did way back in Chapter 3 when you were just controlling the gate voltage of the MOSFET.

The sketch is very similar to the sketch we used to control the brightness of an LED from an Arduino in Chapter 6.

```
// mosfet_motor_speed
int voltsInPin = 0;
int motorPin = 5;

void setup()
{
  pinMode(motorPin, OUTPUT);
}

void loop()
{
  int rawReading = analogRead(voltsInPin);
  int power = rawReading / 4;
  analogWrite(motorPin, power);
}
```

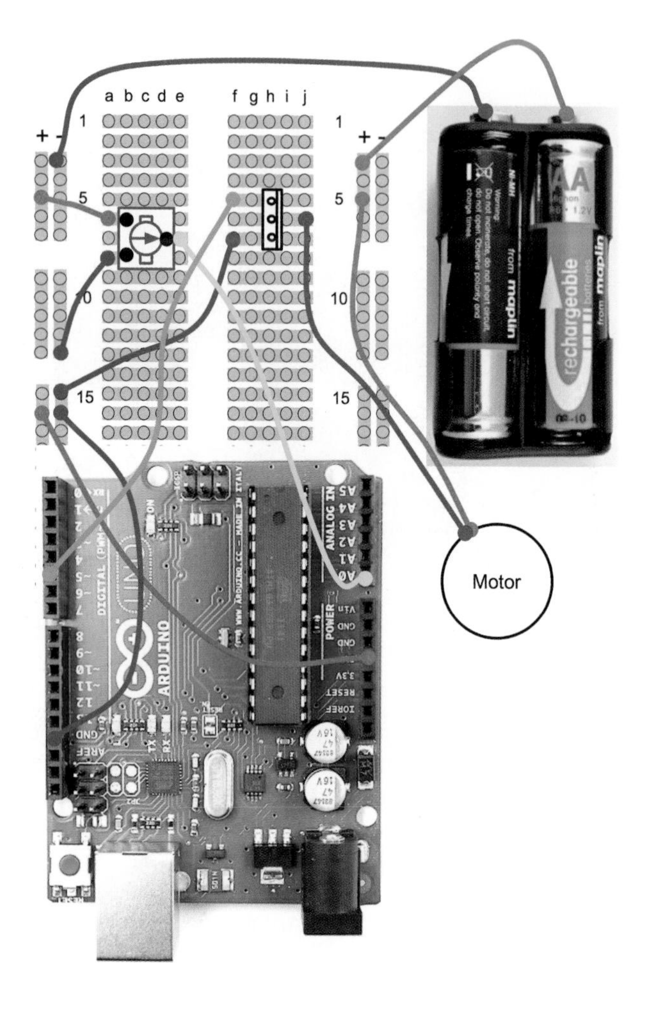

FIGURE 7-19 Breadboard layout for the MOSFET motor control

In the "loop" function, the raw reading of between 0 and 1023 from the analog input is divided by 4 to give us a number between 0 and 255 that is suitable for use with "analogWrite".

How to Control DC Motors with an H-Bridge Module

In the earlier section of this chapter, "How to Control Motor Speed with a Power MOSFET," we saw how you can use a MOSFET to control the speed of a motor. This is fine as long as you always want the motor to turn in the same direction. If you want to be able to reverse the direction of the motor, you need to use something called an H-Bridge.

To change the direction in which a motor turns, you have to reverse the direction in which the current flows. To do this requires four switches or transistors. Figure 7-20 shows how this works, using switches in an arrangement. You can now see why it is called an H-Bridge.

In Figure 7-20, S1 and S4 are closed, while S2 and S3 are open. This allows current to flow through the motor with terminal "A" positive and terminal "B" negative. If we were to reverse the switches so that S2 and S3 are closed and S1 and S4 are open, then "B" will be positive and "A" will be negative, and the motor will turn in the opposite direction.

You may, however, have spotted a danger with this circuit. That is, if by some chance S1 and S2 are both closed, then the positive supply will be directly connected to the negative supply and we will have a short circuit. The same is true if S3 and S4 are both closed at the same time.

You can build an H-Bridge yourself using transistors, and Figure 7-21 shows a typical H-Bridge schematic.

This schematic requires some six transistors and a good few other components. If you wanted to control two motors, you would need some 12 transistors, which causes everything to become quite complicated.

Fortunately, help is on hand as there are several H-Bridge ICs available that usually have two H-Bridges on a single chip and make controlling motors very easy. One such chip is

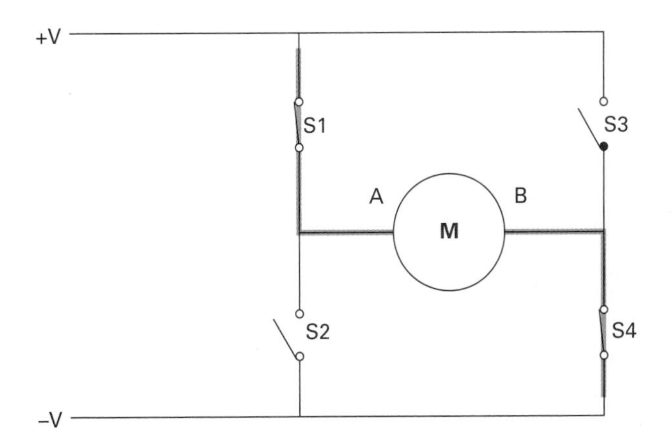

FIGURE 7-20 An H-Bridge using switches

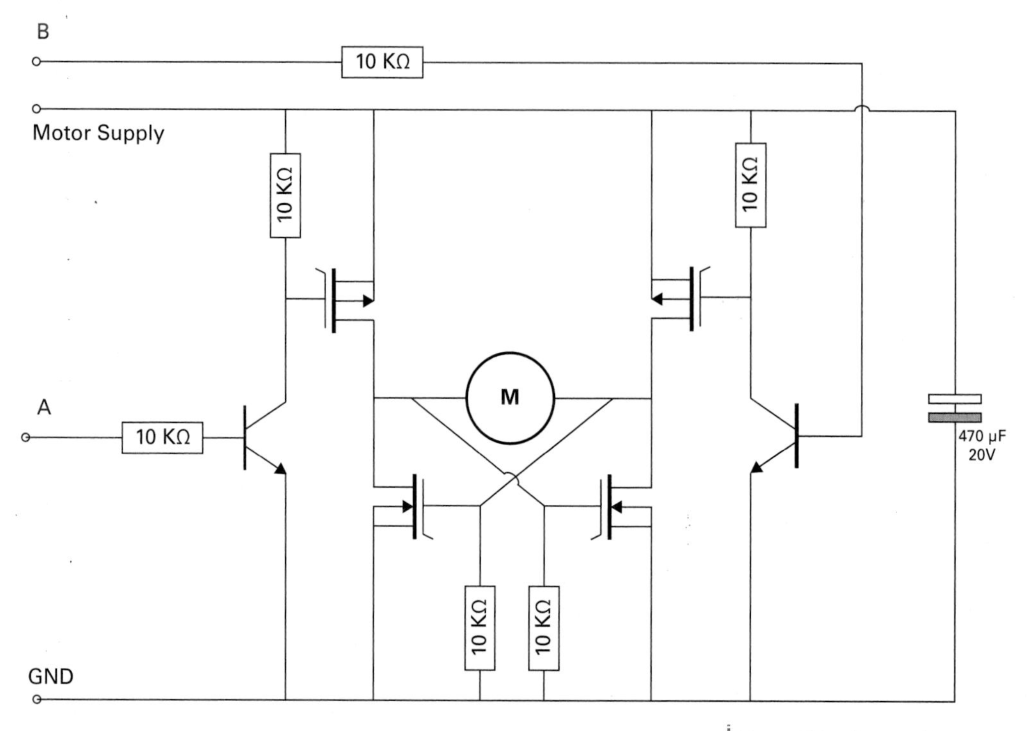

Figure 7-21 An example schematic for an H-Bridge

available as a module from SparkFun (Figure 7-22). You will find similar modules available from other module suppliers.

Figure 7-22 actually shows two of these modules so you can see both sides. The modules are supplied without connectors and the module on the left has pin headers soldered to it. This makes it very easy to use with breadboard.

Table 7-2 shows the pins of this module and explains the purpose of each. The module has two motor channels called A

Figure 7-22 A SparkFun H-Bridge module

Pin Name	Purpose	Purpose	Pin Name
PWMA	PWM input for Channel A	Motor supply voltage (VCC to 15V)	VM
AIN2	Control input 2 for A; high for counter-clockwise	Logic supply (2.7 to 5.5V); only requires 2mA	VCC
AIN1	Control input 1 for A; high for clockwise		GND
STBY	To connect to GND to put the device into "standby" mode.	Motor A connection 1	A01
BIN1	Control input 1 for B; high for clockwise	Motor A connection 2	A02
BIN2	Control input 2 for B; high for counterclockwise	Motor B connection 2	B02
PWMB	PWM input for Channel A	Motor B connection 1	B01
GND			GND

TABLE 7-2 The SparkFun TB6612FNG Breakout Board Pinout

and B and can drive motors with a current of 1.2A per channel with peak currents of over twice that.

We will experiment with this module using just one of its two H-Bridge channels (Figure 7-23).

FIGURE 7-23 Experimenting with the SparkFun TB6612FNG breakout board

You Will Need

To build this, you will need the following items.

Quantity	Name	Item	Appendix Code
1		Solderless breadboard	T5
		Solid-core jumper wire	T6
1		4 × AA battery holder	H1
1		4 × AA batteries	
1		Battery clip	H2
1		LED	K1
1		SparkFun TB6612FNG Breakout Board	M9
1		6V DC motor or gear motor	H6
1		Header pins	K1, H4

The DC motor can be any small motor around 6V.

Breadboard

Before fitting the module onto the breadboard, you need to solder the header pins into place as shown in Figure 7-22. We won't use the bottom two GND connections, so you can just solder the top seven pins on each side.

Figure 7-24 shows the schematic diagram for the experiment, while Figure 7-25 displays the breadboard layout.

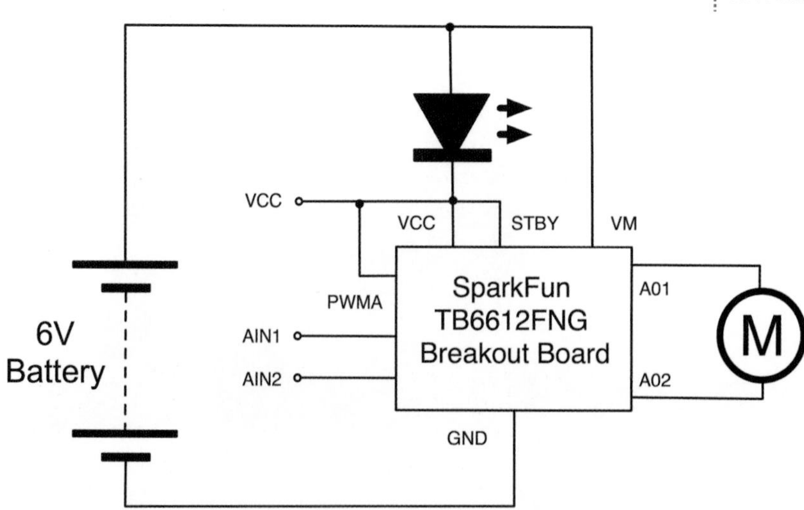

Figure 7-24 Schematic diagram for H-Bridge experiment

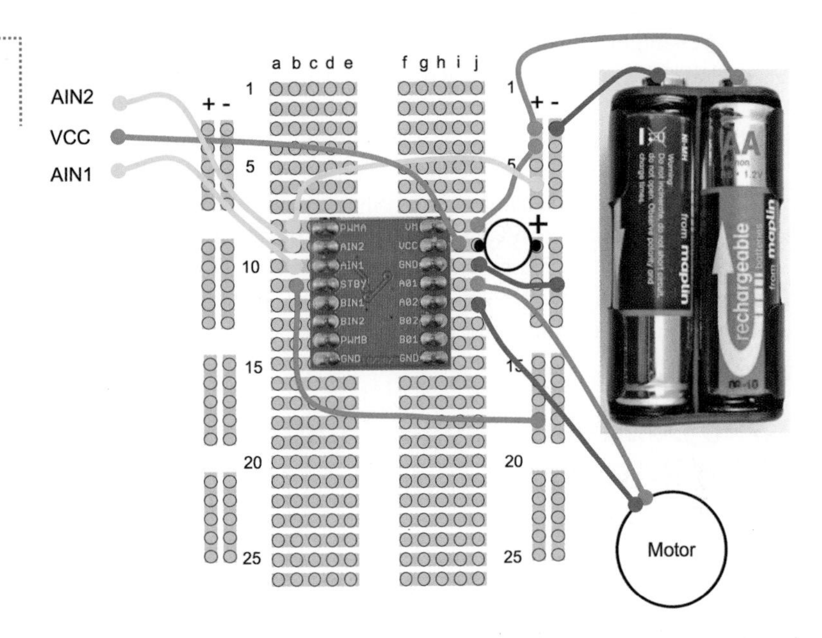

FIGURE 7-25 Breadboard layout for the H-Bridge experiment

The 6V battery pack is actually a slightly higher voltage than is (strictly speaking) allowed for VCC on the module. You would probably get away with the extra half volt above the nominal maximum voltage of 5.5V, but to play it safe, we can use an LED to drop 2V, so that VCC will be around 4V, which is well within its range.

This is a useful trick, but only use it when the current flowing is less than the maximum forward current of the LED. In fact, in this experiment, the current required for VCC is not even enough to make the LED glow.

The PWMA pin is connected to VCC, which simulates the PWN control signal being on all the time—in other words, there is full power to the motor.

Next, put everything on the breadboard as shown in Figure 7-25.

Using the Control Pins

Three of the leads from the breadboard do not actually go anywhere. You will control the motor, touching the red lead going to VCC to AIN1, and then to AIN2, in turn. Note how the motor turns first in one direction and then the other.

You might be wondering why there are two control pins, as well as the PWM pin for each motor channel. In theory, you could have one direction pin and one PWM pin, and if the PWM power was zero, then the motor would not turn at all.

The reason we have three pins to control each motor (PWM, IN1, and IN2) rather than just two is that if both IN1 and IN2 are high (connected to VCC), then the H-Bridge operates in a "braking" mode, which provides electrical braking of the motor, slowing it down. This feature is not often used, but can be useful if you want to stop the motor quickly.

How to Control a Stepper Motor with an H-Bridge Module

Normal DC motors are nice and easy to use. There are just two connections to make and if the voltage is applied one way it turns clockwise; reverse the polarity and it turns counterclockwise. The down side to normal DC motors is that if you want to know what position it has turned to, you have to use some kind of sensor.

Stepper motors are entirely different kinds of motors. They commonly have four connections. Figure 7-26 shows how a stepper motor works. Or more specifically, a bipolar stepper motor, which is the one we will try out.

The motor contains a toothed rotor where each of the teeth of the rotor are magnets, of alternating north and south poles.

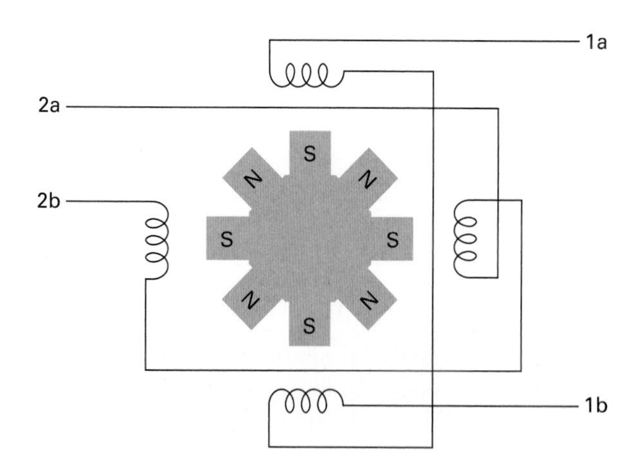

FIGURE 7-26 How a bipolar stepper motor works

Four coils, acting as electromagnets will, when energized in the right order, move the rotor around one step. The coils are arranged in pairs, wired so that as one pushes, its opposite number pulls.

Most stepper motors will have far more steps than the eight shown in Figure 7-26, sometimes 200 or more. This makes the motors very flexible, because they can run freely just like any other motor, by sending the stepping pulses quickly, or they can be controlled very precisely by just moving them forward one step at a time. For this reason, you will find stepper motors in inkjet printers and also 3D printers.

Because the stepper motor will only turn, if we generate a series of pulses in the right order, and need to be able to reverse the direction of current flow in the coils, we can use an Arduino to generate the control signals and an H-Bridge module to supply the power to the coils (Figure 7-27).

Figure 7-28 shows the schematic diagram for this arrangement.

Identifying which lead is which on a stepper motor sometimes requires a bit of trial and error. Using a multimeter, you can measure the resistance between pairs of leads and therefore work out which leads are connected to the same coil.

Another way of finding the leads that belong to the same coil is to hold two of the leads together and see if it makes the shaft of the motor more difficult to turn. Strange, but true!

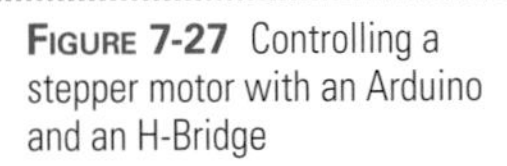

FIGURE 7-27 Controlling a stepper motor with an Arduino and an H-Bridge

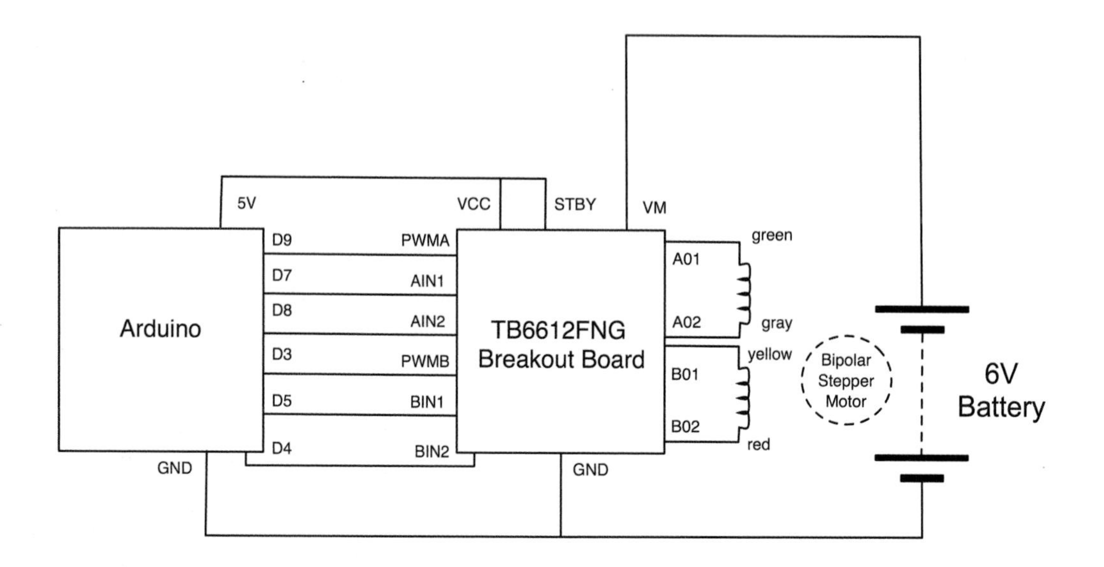

FIGURE 7-28 Schematic diagram for a stepper motor control

If when you turn it on, the motor doesn't turn, you will only need to swap over one of the coil's leads. The colors indicated in Figure 7-28 match the lead colors for the Adafruit motor.

Although the motors suggested are 12V, they will still work using the 6V supplied by the battery pack. However, do not try and power them from the 5V supply of the Arduino. They draw too much current.

You Will Need

To build this, you will need the following items.

Quantity	Item	Appendix Code
1	Solderless breadboard	T5
	Solid-core jumper wire	T6
1	6 × AA battery holder	H8
1	6 × AA batteries	
1	TB6612FNG breakout board	M9
1	Bipolar stepper motor	H13
1	Arduino Uno/Leonardo	M2/M21
1	USB lead; Type B for Uno, Micro USB for Leonardo	

Construction

Figure 7-29 shows the breadboard layout.

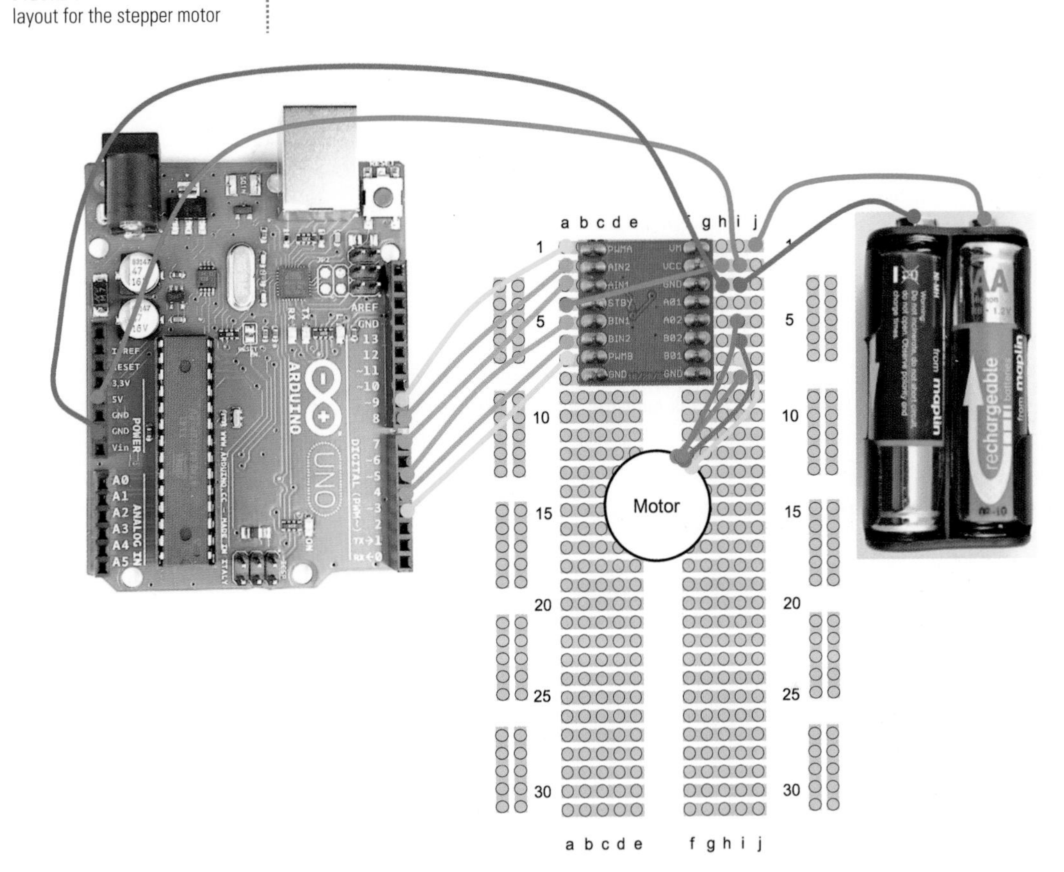

FIGURE 7-29 Breadboard layout for the stepper motor

Software

The example sketch ("stepper") will first of all turn the motor in one direction through 200 steps and then pause for a second before turning it in the opposite direction through 200 steps. For a 200-step motor, each turn will be a full 360 degrees.

First, the pin variables are defined and the "setup" function sets them all to be outputs.

```
// stepper

int PWMApin = 9;
int AIN1pin = 7;
```

```
int AIN2pin = 8;
int PWMBpin = 3;
int BIN1pin = 5;
int BIN2pin = 4;

void setup()
{
  pinMode(PWMApin, OUTPUT);
  pinMode(AIN1pin, OUTPUT);
  pinMode(AIN2pin, OUTPUT);
  pinMode(PWMBpin, OUTPUT);
  pinMode(BIN1pin, OUTPUT);
  pinMode(BIN2pin, OUTPUT);
}
```

The "loop" function then directs the motor in one direction (forward) for 200 steps, pauses a second, and then directs the motor back the same number of steps, but with half the delay between steps before pausing another second. It will continue this indefinitely.

```
void loop()
{
  forward(10, 200);
  delay(1000);
  back(5, 200);
  delay(1000);
}
```

The functions "forward" and "back" both take two parameters. The first is the delay between each step in milliseconds, and the second is the number of steps to take.

The "forward" and "back" functions use the function "setStep" to set the right polarities of the two coils, in the pattern 1010, 0110, 0101, 1001.

```
void forward(int d, int steps)
{
  for (int i = 0; i < steps / 4; i++)
  {
    setStep(1, 0, 1, 0);
    delay(d);
    setStep(0, 1, 1, 0);
    delay(d);
```

```
    setStep(0, 1, 0, 1);
    delay(d);
    setStep(1, 0, 0, 1);
    delay(d);
  }
}
```

To make the motor turn in the opposite direction, the pattern is just reversed.

```
void back(int d, int steps)
{
  for (int i = 0; i < steps / 4; i++)
  {
    setStep(1, 0, 0, 1);
    delay(d);
    setStep(0, 1, 0, 1);
    delay(d);
    setStep(0, 1, 1, 0);
    delay(d);
    setStep(1, 0, 1, 0);
    delay(d);
  }
}
```

The "setStep" function actually sets the appropriate outputs of the motor controller.

```
void setStep(int w1, int w2, int w3, int w4)
{
  digitalWrite(AIN1pin, w1);
  digitalWrite(AIN2pin, w2);
  digitalWrite(PWMApin, 1);
  digitalWrite(BIN1pin, w3);
  digitalWrite(BIN2pin, w4);
  digitalWrite(PWMBpin, 1);
}
```

How to Make a Simple Robot Rover

In this project, we will create a little roving robot. To do this, we will use the RF remote control we used in the section "How to Use a Wireless Remote Module," with the H-Bridge module we

FIGURE 7-30 The robot rover

just discussed in the section "How to Control DC Motors with an H-Bridge Module," along with an Arduino.

The project will demonstrate how to use an Arduino to control a motor module.

The robot (Figure 7-30) will be built using a low-cost robot chassis kit that includes two gear motors.

The robot is built using a small breadboard that holds both the motor module and the RF receiver module. So apart from putting pin headers on the motor controller, there is no soldering to be done in this project.

You Will Need

To build this, you will need the following items.

Quantity	Name	Item	Appendix Code
1		Small solderless breadboard	H12
		Solid-core jumper wire	T6
1		6 × AA battery holder	H8
1		6 × AA batteries	
*1		Battery clip to 2.1mm DC jack adapter	H9

Quantity	Name	Item	Appendix Code
1		LED	K1
1		SparkFun TB6612FNG breakout board	M9
1		Magician chassis	H7
1		Header pins	K1, H4
1	C1	1000μF 16V capacitor	C1
1	C2	100μF 16V capacitor	C2
1		Arduino Uno/Leonardo	M2/M21
1		USB lead; Type B for Uno, Micro USB for Leonardo	

* If you use the Adafruit battery box that is already terminated in a 2.1mm plug, then you do not need this.

Construction

Figure 7-31 shows the schematic diagram for the rover.

The use of modules simplifies this greatly. The only additional components that have been added are the two capacitors: C1 and C2. These are necessary to prevent sudden drops in battery voltage (as the motors start) from causing the Arduino to reset.

FIGURE 7-31 Schematic diagram for the rover

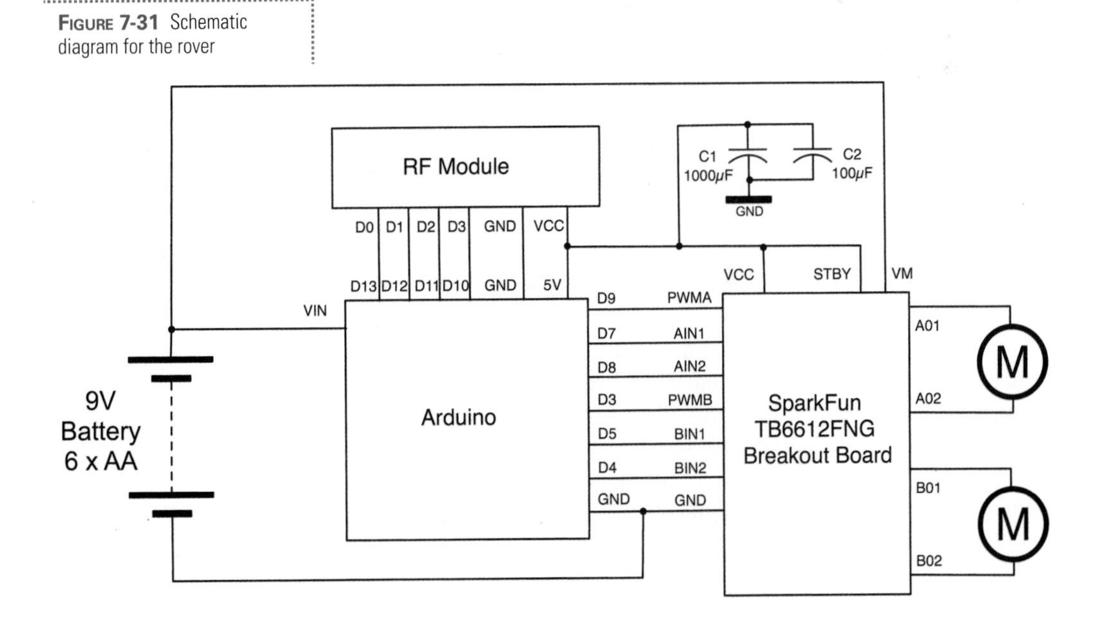

Step 1. Construct the Magician Chassis

The project is built around the Magician Chassis (Figure 7-32). This comes as a kit that is fixed together with nuts and bolts. Follow the instructions that come with the Chassis, but do not attach the battery box that comes with the kit or the support pillar that is right in the middle of the board. This is because, to power the Arduino, you need a bit more than the 5V to 6V that four AAs can supply. So, you are going to replace the battery box with one that takes six AA batteries rather than just four.

FIGURE 7-32 The Magician Chassis

Step 2. Program the Arduino

It is a good idea to program the Arduino with the sketch before you start attaching the electronics. Load the sketch "rover" onto the Arduino.

Step 3. Attach the Arduino and Breadboard

Find suitable mounting holes on the chassis and use small nuts and bolts to attach the Arduino. You can also use an elastic band for this. Some breadboards come with a self-adhesive backing, and you can use this to attach it to the chassis. For a less permanent way of attaching it, a rubber band will work just fine.

Step 4. Build the Breadboard

Figure 7-33 shows the breadboard layout for the project and how it is wired up to the Arduino.

There are quite a lot of wires on this project, so check all the connections once you think you have finished. Photocopying the page of the book and checking off with a pencil each connection as it's made is a good way of ensuring they are all there.

Also, not unlike our normal large breadboard, when wiring up this breadboard we have used the outer supply rail as GND and the inner rail as 5V.

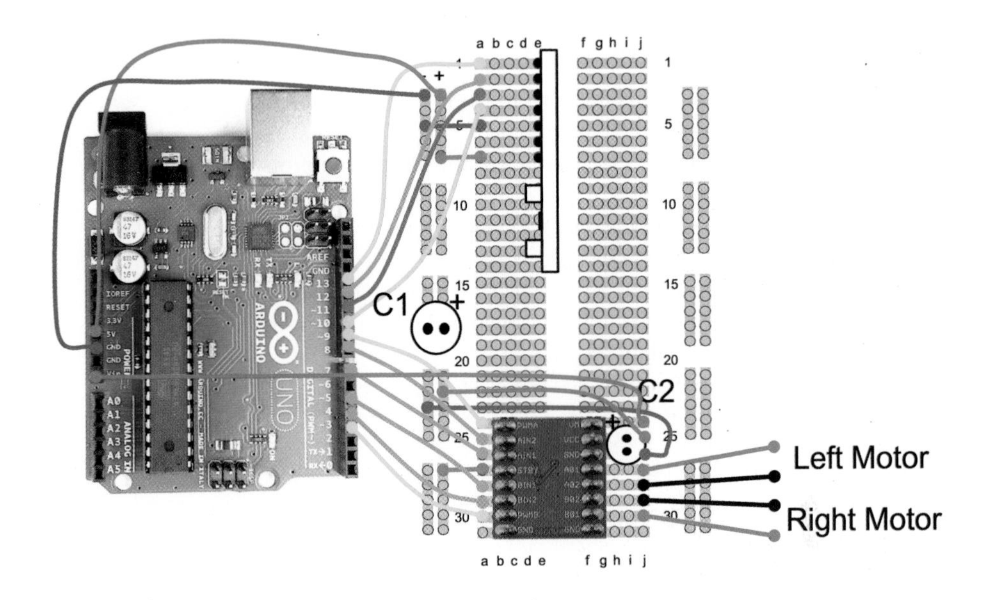

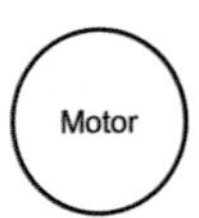

FIGURE 7-33 Breadboard layout for the rover

Step 5. Wire up the Motors

Each motor has a red and a black lead. So find the leads going to the left motor and attach them to the breadboard rows connected to the A01 and A02 connections of the motor module. Then, do the same for the right motor, connecting them to B01 and B02.

Step 6. Attach the Battery

If the battery box is made up of two rows of batteries, it will be quite a snug fit and the top surface of the chassis will need to bend out a little to accommodate it. If it is all in one row, like the Adafruit box, then you can attach it to the bottom layer of the chassis using small nuts and bolts.

Testing

When everything is assembled and ready to go, attach the battery and try the project out by pressing the buttons on the remote. The C button will set the robot running forwards, the B button will make it rotate to the right on the spot, and D will make it turn to the left. The A button will bring the robot to a halt.

Software

The sketch for this project is too long to list in full here, so we will just look at some of the main points.

The RF receiver toggles the output for the button you press. So press it once and it turns on, press it again and it turns off. However, this is not really the way we want it to work. We just want to know when a button has been pressed.

To do that, we keep track of the last state of each of the outputs, and only when the output has changed do we report the change. This uses the following array to store the output states and another array "remotePins":

```
int remotePins[] = {10, 11, 12, 13};
int lastPinStates[] = {0, 0, 0, 0};
```

The function that detects that a change has occurred is as follows:

```
int getKeyPress()
{
  // the outputs on the RF module toggle
  // so see what's changed and that's the
  // key that was pressed
  int result = -1;
  for (int i = 0; i < 4; i++)
  {
    int remoteInput = digitalRead(remotePins[i]);
    //Serial.print(remoteInput);
    if (remoteInput != lastPinStates[i])
    {
      result = i;
    }
    lastPinStates[i] = remoteInput;
  }
  return result;
}
```

The main loop calls this function to detect any key presses and then calls the appropriate function for the button.

```
void loop()
{
   int keyPressed = getKeyPress();
   Serial.println(keyPressed);
```

```
    if (keyPressed == 3)
    {
      stopMotors();
    }
    else if (keyPressed == 0)
    {
      turnLeft();
    }
    else if (keyPressed == 2)
    {
      turnRight();
    }
    else if (keyPressed == 1)
    {
      forward();
    }
    delay(20);
}
```

The functions that control the movement are all very similar. The function for turning left is shown next.

```
void turnLeft()
{
  digitalWrite(AIN1pin, HIGH);
  digitalWrite(AIN2pin, LOW);
  analogWrite(PWMApin, slowPower);
  digitalWrite(BIN1pin, LOW);
  digitalWrite(BIN2pin, HIGH);
  analogWrite(PWMBpin, slowPower);
}
```

This sets the AIN and BIN pins—in this case, to set the motors turning in opposite directions. The PWM power is controlled by a call to "analogWrite" using one of two values held in variables ("fullPower" and "slowPower").

How to Use a Seven-Segment LED Display Module

Seven-segment LED displays have a nice retro feel to them.

LED displays made up of a number of LEDs contained in a single package can be a challenge to control. Such displays will normally be controlled using a microcontroller; however, it is not necessary to use a microcontroller output pin to each individual LED. But rather, multi-LED displays are organized as "common anode" or "common cathode," with all the LED terminals of the anode or cathode connected together and then brought out through one pin. Figure 7-34 shows how a common cathode seven-segment display might be wired internally.

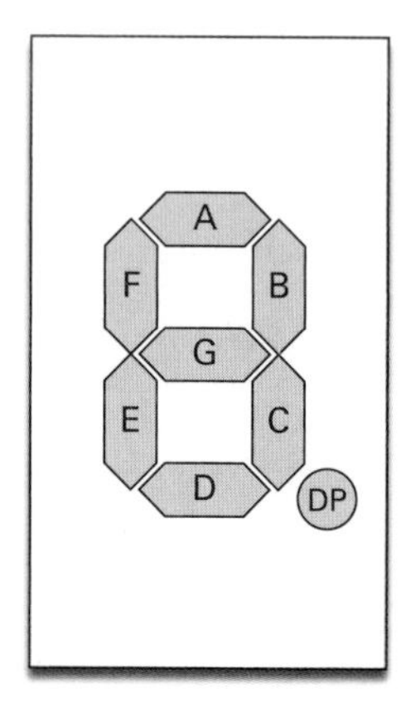

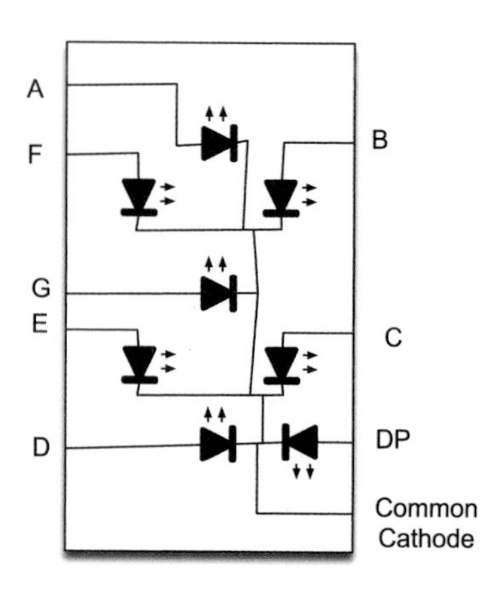

FIGURE 7-34 A common cathode LED display

In a common cathode display like this, the common cathode would be connected to ground and each segment anode driven by a microcontroller pin through a separate resistor. Do not be tempted to use one resistor on the common pin, and don't use any resistors on the non-common connections, since the current will be limited no matter how many LEDs are lit. Because of this, the display will get dimmer the more LEDs are illuminated.

It is quite common for multiple displays to be contained in the same case—for example, the three-digit, seven-segment common cathode LED display shown in Figure 7-35.

In this kind of display, each digit of the display is like the single-digit display of Figure 7-35, and has its own common cathode. In addition, all the A segment anodes are connected together, as are each segment.

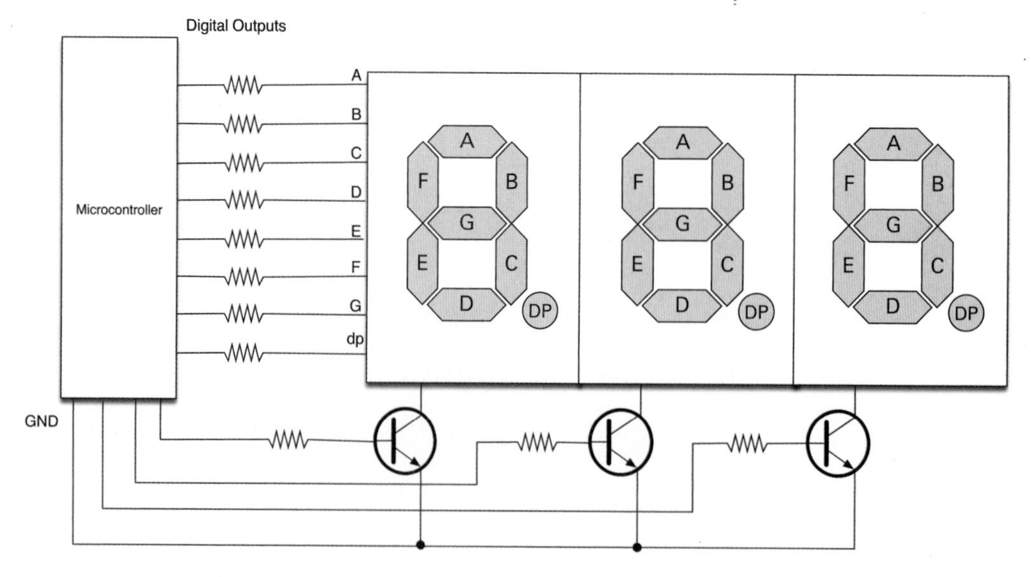

FIGURE 7-35 A three-digit, seven-segment LED display

FIGURE 7-36 A four-digit, seven-segment I2C display

The Arduino using the display will then activate each common cathode in turn and then turn on the appropriate segments for that digit, and then move onto the next digit, and so on. This refresh happens very quickly so that the display appears to display different numbers on each digit. This is called multiplexing.

Note the use of transistors to control the common cathodes. This is simply to handle the current of potentially eight LEDs at once, which would be too much for most microcontrollers.

Fortunately for us, there is a much simpler way to use multi-digit, seven-segment LED displays. Modules ride to the rescue once again!

Figure 7-36 shows a four-digit, seven-segment LED display that has just four pins on its connector, and two of them are for power.

You Will Need

To build this, you will need the following items.

Quantity	Item	Appendix Code
1	Solderless breadboard	T5
	Solid-core jumper wire	T6
1	Arduino Uno/Leonardo	M2/M21
1	USB lead; Type B for Uno, Micro USB for Leonardo	
1	Adafruit seven-segment display w/I2C backpack	M19

Construction

The module comes as a kit, so start by following the instructions that accompany the module to assemble it.

The LED module uses a type of serial interface on the Arduino called I2C (pronounced "I squared C"). This requires just two pins, but they have to be the two pins above "AREF" on the Arduino Uno. These pins are named SDA and SCL.

This means that, frustratingly, the module will not just plug straight into the Arduino, we will need to use breadboard.

Figure 7-37 shows the breadboard layout and Figure 7-38 the breadboard itself, with the seven-segment display in action.

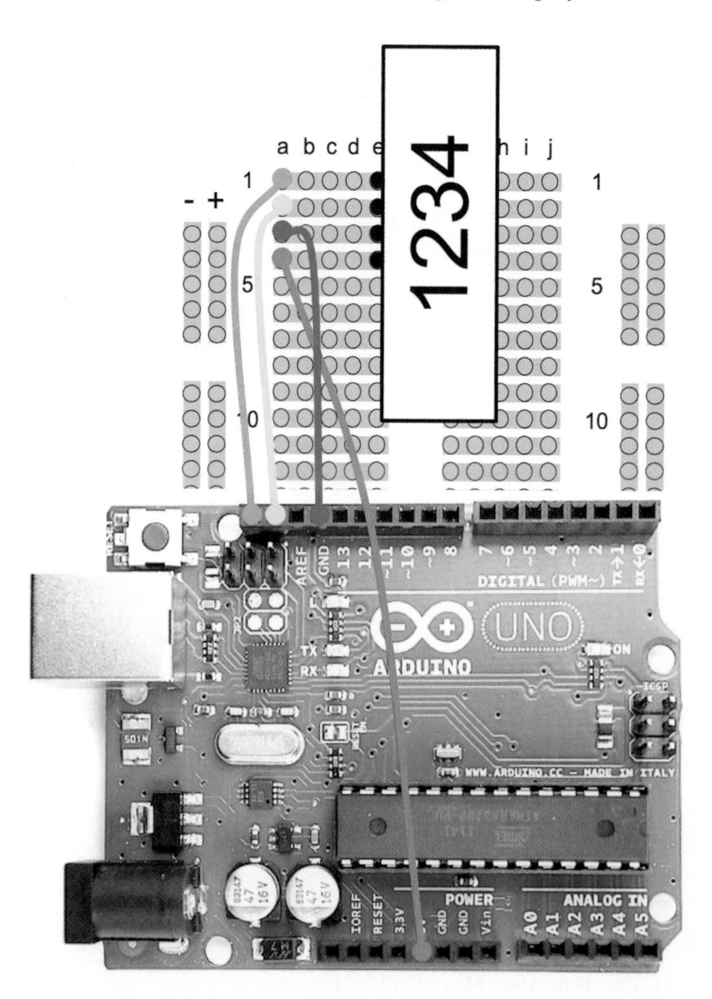

FIGURE 7-37 Breadboard layout for using the seven-segment display

FIGURE 7-38 The seven-segment display in action

Software

Adafruit provides a library to simplify the use of the module. You need to download this and copy the library folder into the "libraries" folder in your Arduino documents folder. See the instructions on Adafruit's web site at www.adafruit.com/products/880.

The three libraries that the module requires are loaded using the #includes statements.

```
// seven_seg_display

#include <Wire.h>
#include "Adafruit_LEDBackpack.h"
#include "Adafruit_GFX.h"
```

The following line assigns a variable to the display object so we can tell it what to display.

```
Adafruit_7segment disp = Adafruit_7segment();
```

The "setup" function begins serial communication on the I2C pins and then initializes the display. The value 0x70 is the I2C address of the display module. This is the default value for its address, but there are solder connections on the module you can short together to change the address. You might want to do this if you need to use more than one display, since each display must have a different address.

```
void setup()
{
  Wire.begin();
  disp.begin(0x70);
}
```

The "loop" function simply displays the current number of milliseconds since the board was reset, divided by 10. The display will therefore count up in 1/100ths of a second.

```
void loop()
{
  disp.print(millis() / 10);
  disp.writeDisplay();
  delay(10);
}
```

How to Use a Real-Time Clock Module

You could write an Arduino sketch to keep track of the time, but as soon as you unplugged it, it would forget the time. The way around this problem is to use an RTC (real-time clock) like the one shown in Figure 7-39.

FIGURE 7-39 An RTC module

FIGURE 7-40 A "simple" digital clock

This particular module is also an Adafruit product. There are lots of similar modules out there, but their pin allocations may be different.

The RTC includes a lithium battery that will last for years, and provides enough power to keep the correct time when the module is not powered.

We can combine the RTC module with the seven-segment display module we used previously and make ourselves a simple digital clock (Figure 7-40).

You Will Need

To build this, you will need the following items.

Quantity	Item	Appendix Code
1	Solderless breadboard	T5
	Solid-core jumper wire	T6
1	Arduino Uno/Leonardo	M2/M21
1	USB lead; Type B for Uno, Micro USB for Leonardo	
1	Adafruit seven-segment display w/I2C backpack	M19
1	RTC module	

Construction

The RTC module also comes as a kit, so start by following the instructions that accompany the module to assemble it.

The RTC module also uses I2C and has a different address to the display, so we do not need to change anything.

Figure 7-41 shows the breadboard layout for the clock.

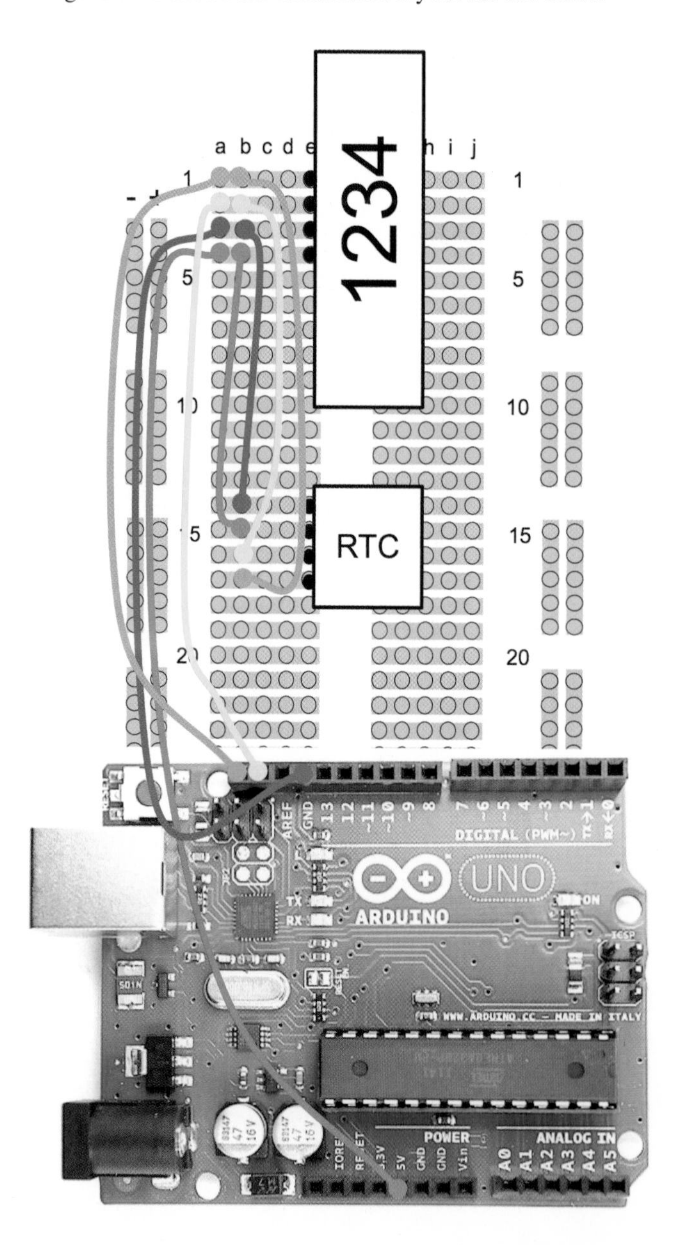

FIGURE 7-41 Breadboard layout for the clock

Software

Load up the sketch "clock" onto your Arduino. The display should immediately start showing the time your computer is set to.

Much of this sketch is the same as that in the section "How to Use a Seven-Segment LED Display Module." But there is one additional library for the RTC module that we need to import. Instructions for downloading this are linked from the product page for the RTC module (www.adafruit.com/products/264).

```
// clock

#include <Wire.h>
#include "Adafruit_LEDBackpack.h"
#include "Adafruit_GFX.h"
#include "RTClib.h"
```

In addition to creating a display to use, we now have to give the RTC a name. Let's call it "RTC".

```
RTC_DS1307 RTC;
Adafruit_7segment disp = Adafruit_7segment();
```

The "setup" function now has an additional command to start the RTC so it is ready to receive commands. The "if" statement checks to see if the clock part of the RTC is active. If this is the first time it has been used, it will not be, so if this is the case, it initializes it to the programming computer's time.

```
void setup()
{
  Wire.begin();
  RTC.begin();
  if (! RTC.isrunning())
  {
    RTC.adjust(DateTime(__DATE__, __TIME__));
  }
  disp.begin(0x70);
}
```

The main loop now reads the time from the RTC and displays it. It also uses the display libraries' "drawColon" function to make the colon flash by turning it on and off with a half-second delay in between.

```
void loop()
{
  disp.print(getDecimalTime());
  disp.drawColon(true);
  disp.writeDisplay();
  delay(500);
  disp.drawColon(false);
  disp.writeDisplay();
  delay(500);
}
```

The "getDecimalTime" function reads the hours and minutes from the RTC and turns them into a decimal number that can be written to the display. The first two digits will contain the hour, and the left two digits the minute.

```
int getDecimalTime()
{
  DateTime now = RTC.now();
  int decimalTime = now.hour() * 100 + now.minute();
  return decimalTime;
}
```

Summary

In addition to the modules here, you will find lots of other useful modules on the web sites of companies like Adafruit and SparkFun. The web sites also include some information on how to use the modules and their specifications. If you find a module you would like to make use of, the first step is to research how you could use it. As well as the datasheets and tutorial information on the supplier's web site, you will often find instructions on building the projects if you search for the module on the Internet.

8

Hacking with Sensors

Chapters 6, 7, and 8 all overlap somewhat, as many sensors are also modules and both can often be used with an Arduino.

In this chapter, we will look at how to use a range of sensors, whether with a little supporting electronics or as an input to an Arduino, or sometimes both.

How to Detect Noxious Gas

In this section, you will use a methane sensor (Figure 8-1).

While they look like they should be expensive, these sensors are really quite low cost. They include a small heater (connected between the two H connections) and a catalytic sensing element whose resistance changes depending on the concentration of methane. Although the project will run on batteries, it will burn through them pretty quickly because these sensors have a heating element that will consume 150 to 200 mA.

The sensing of methane does have lots of sensible scientific and industrial uses. However, we will use this technological know-how for the puerile activity of … detecting farts.

You Will Need

To experiment with this gas sensor, you will need the following items.

Quantity	Names	Item	Appendix Code
1	D1	LED	K1
1	R1	10kΩ trimpot	K1
1	R2	10kΩ resistor	K2
1	R3	470Ω resistor	K2
1	IC1	LM311 comparator	S7

Quantity	Item	Appendix Code
1	Methane sensor MQ-4	M11
1	Piezo buzzer (with own oscillator)	M10
1	Solderless breadboard	T5
	Solid-core jumper wire	T6
1	4 × AA battery holder	H1
1	4 × AA batteries	
1	Battery clip	H2
1	* Arduino Uno/Leonardo	M2/M21
1	* USB lead; Type B for Uno, Micro USB for Leonardo	

* Only required if you want to connect the detector to an Arduino.

The piezo sounder must be of the type that includes its own oscillator circuit and will work at 6V.

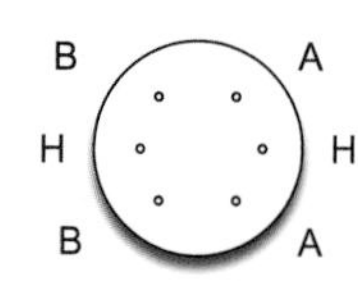

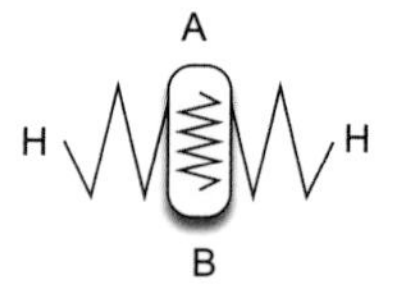

FIGURE 8-1 A methane sensor

The LM311 Comparator

Figure 8-2 shows the schematic diagram for the gas detector.

The key to this circuit is the comparator IC (LM311). Comparators, as the name suggests, compare voltages. If the voltage at its "+" connection is greater than the voltage at its "–"

FIGURE 8-2 Schematic diagram for the gas detector

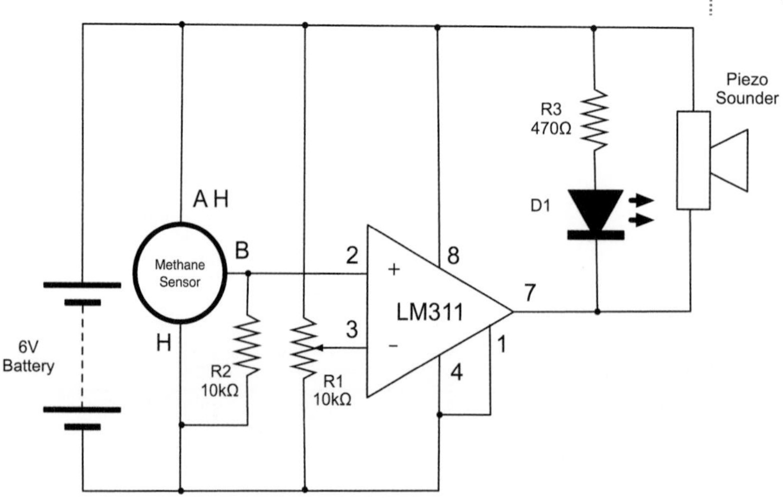

connection, then its output turns on. In this case, that will light the LED and sound a buzzer.

The trimpot supplies a threshold voltage to the negative input of the comparator. To use the gas detector, the trimpot is turned until the LED just goes out. It will come back on if the output from the sensor increases enough to exceed the value at the comparator's negative input.

FIGURE 8-3 Attaching leads to the sensor

The sensor has rather unusual connections. It has six connections, but some of them are doubled up and connected behind the scenes (see Figure 8-1). The H connectors supply a heating element that warms the catalyst layer between A and B. When methane is detected, the resistance between A and B falls. R2 forms a voltage divider with the sensing element. One benefit of the sensor basically being two resistors—one acting as a heater and the other as a sensor—is that the pin connections are reversible.

The sensor leads are thick and at a strange spacing, so they will not fit in breadboard. For this reason, we solder some leads to them (Figure 8-3).

Rather than solder wires to all the leads, we can just solder the following connections:

- A red positive supply lead to all the pins on one side of the sensor (the two A pins and one H connection)
- The resistor R2 between B and the GND side of the heater
- A GND lead to the GND side of the heater (black)
- An output lead to B (yellow)

Breadboard

Figure 8-4 shows the breadboard layout for the gas detector, while Figure 8-5 displays the project itself.

The breadboard layout is very straightforward, but do make sure that the IC is the correct way around. When it is all assembled, I will leave you to find your own way of testing it. Just a note that breathing on the sensor will also set it off.

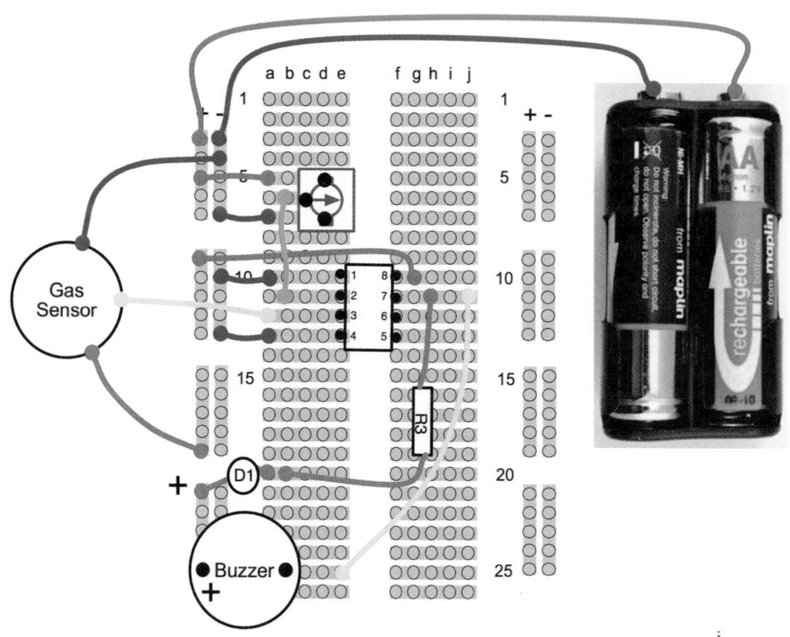

FIGURE 8-4 Breadboard layout for the gas detector

Using a Gas Sensor with an Arduino

In soldering the three leads onto the methane sensor, we have also made something that we can easily attach directly to an Arduino (Figure 8-6).

Connect the positive supply connection from the sensor to 5V on the Arduino, GND on the sensor to GND on the Arduino, and the output of the sensor to A3.

Since this sensor can use up to 200 mA, you must power it from the real 5V and GND connections on the Arduino and not use the trick of powering it from a digital output.

The following sketch ("methane") prints the readings from the sensor into the Serial Monitor. Again, note that if you breathe on the sensor, the reading will increase.

FIGURE 8-5 The gas detector

FIGURE 8-6 Using the gas sensor with an Arduino

```
// methane

int analogPin = 3;

void setup()
{
  Serial.begin(9600);
  Serial.println("Methane Detector");
}

void loop()
{
  Serial.println(analogRead(analogPin));
  delay(500);
}
```

How to Measure Something's Color

The TCS3200 is a handy little IC for measuring the color of something. There are several different variations on this chip, but they all work the same way. The chip has a transparent case, and dotting its surface are photodiodes with different color filters over them (red, green, and blue). You can read the relative amounts of each primary color.

The easiest way to use the chip is to buy a module like the one shown in Figure 8-7.

This module, which cost less than USD 10, also has four white LEDs that illuminate the object whose color you want to measure, as well as convenient header pins.

Figure 8-7 A light-sensing module

Table 8-1 shows the connections on the module and their purpose. With the exception of the power to the LEDs, these connections are taken straight from the IC, so any module you find that uses the TCS3200 is likely to have the same connections, even if they are not quite in the same place.

The IC does not produce an analog output, but instead varies the frequency of a train of pulses. You choose which color the pulse frequency corresponds to by changing the values on the digital inputs S2 and S3.

You Will Need

Quantity	Item	Appendix Code
1	Arduino Uno/Leonardo	M2/M21
1	USB lead; Type B for Uno, Micro USB for Leonardo	
1	Color-sensing module	M12
1	Male-to-female jumper set	T12

<table>
<tr><th>Pin</th><th>Description</th><th>Description</th><th>Pin</th></tr>
<tr><td>S0</td><td rowspan="2">S0 and S1 select the frequency range. Both should be set HIGH.</td><td>2.5V to 5.5V</td><td>VCC</td></tr>
<tr><td>S1</td><td>Ground</td><td>GND</td></tr>
<tr><td>S2</td><td rowspan="2">Red—S2 and S3 LOW
Green—S2 and S3 HIGH
Blue—S2 LOW, S3 HIGH
White—S2 HIGH, S3 LOW</td><td>Output Enable—set to LOW to effectively turn the chip on.</td><td>OE</td></tr>
<tr><td>S3</td><td>Tie to ground with the attached jumper to turn the LEDs on.</td><td>LED</td></tr>
<tr><td>OUT</td><td colspan="2">The output pulses.</td><td>GND</td></tr>
</table>

TABLE 8-1 Color-Sensing Module Pinout

Construction

Construction is perhaps too strong a word for it. The module will fit directly into the Arduino (Figure 8-8), facing outward. It will make the following connections:

- S0 module to D3 Arduino
- S1 module to D4 Arduino
- S2 module to D5 Arduino
- S3 module to D6 Arduino
- OUT module to D7 Arduino

You will also need three male-to-female jumper leads to connect:

- VCC module to 5V Arduino
- GND module to GND Arduino
- OE module to GND Arduino

Figure 8-9 shows the module sensing colors on a Rubik's cube.

FIGURE 8-8 A light sensor attached to an Arduino

Software

The sketch "color_sensing" demonstrates the use of this module.

```
// color_sensing
int pulsePin = 7;
int prescale0Pin = 3;
int prescale1Pin = 4;
int colorSelect0pin = 5;
int colorSelect1pin = 6;
```

The pins are named according to their function rather than using the module pin names.

The "setup" function sets the appropriate pin modes and then sets both the "prescale" pins that control the

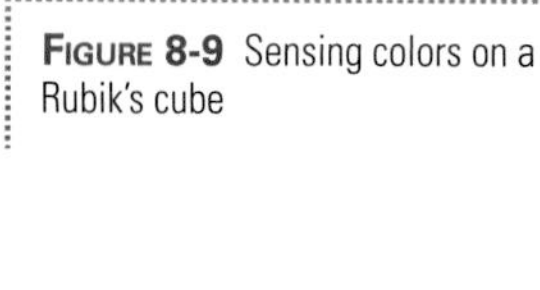

Figure 8-9 Sensing colors on a Rubik's cube

output frequency range to HIGH, starts serial communication, and then displays a welcome message.

```
void setup()
{
  pinMode(prescale0Pin, OUTPUT);
  pinMode(prescale1Pin, OUTPUT);
  // set maximum prescale
  digitalWrite(prescale0Pin, HIGH);
  digitalWrite(prescale1Pin, HIGH);
  pinMode(colorSelect0pin, OUTPUT);
  pinMode(colorSelect1pin, OUTPUT);
  pinMode(pulsePin, INPUT);
  Serial.begin(9600);
  Serial.println("Color Reader");
}
```

The "loop" function reads the three different colors (more on that later) and displays a message depending on the dominant color. Note that the lower the value, the brighter that particular color.

```
void loop()
{
  long red = readRed();
  long green = readGreen();
  long blue = readBlue();
  if (red < green && red < blue)
  {
```

```
    Serial.println("RED");
  }
  if (green < red && green < blue)
  {
    Serial.println("GREEN");
  }
  if (blue < green && blue < red)
  {
    Serial.println("BLUE");
  }
  delay(500);
}
```

Each of the functions—"readRed", "readGreen", "readBlue", and "readWhite"—just call a function "readColor" with the appropriate values for S2 and S3.

```
long readRed()
{
  return (readColor(LOW, LOW));
}
```

The function "readColor" first sets the appropriate pins for the color and records a start time in the variable "start". It then waits for 1000 pulses to happen. Afterward, it returns the difference between the current time and the start time.

```
long readColor(int bit0, int bit1)
{
  digitalWrite(colorSelect0pin, bit0);
  digitalWrite(colorSelect1pin, bit1);
  long start = millis();
  for (int i=0; i< 1000; i++)
  {
    pulseIn(pulsePin, HIGH);
  }
  return (millis() - start);
}
```

Although not actually used, there is also a function in the sketch that writes the color values to the Serial Monitor.

```
void printRGB()
{
  Serial.print(readRed()); Serial.print("\t");
  Serial.print(readGreen()); Serial.print("\t");
```

```
  Serial.print(readBlue()); Serial.print("\t");
  Serial.println(readWhite());
}
```

How to Detect Vibration

Piezo vibration sensors, like the one from SparkFun shown in Figure 8-10, are very easy to use with an Arduino.

The sensors are a thin strip of piezoelectric material with a rivet in the end acting as a weight. When there is a vibration, the weight moves, stressing the piezo material that produces a spike in voltage. Measured with the right test equipment, this spike can be as high as 80V. However, because we are going to connect it to an analog input on an Arduino, the resistance of that input will be sufficient to damp the voltage to a level that will not harm our Arduino.

Figure 8-10 A piezo vibration sensor

You Will Need

To detect vibration with your piezo sensor, you will need the following items.

Quantity	Item	Appendix Code
1	Arduino Uno/Leonardo	M2/M21
1	USB lead; Type B for Uno, Micro USB for Leonardo	
1	Piezo vibration sensor	M13
1	LED	K1
1	220Ω resistor	K2

Construction

The piezo sensor is another very Arduino-friendly sensor. It can be just plugged into the Arduino sockets. In this case, it is plugged into pins A0 and A1. A0 will be set to an output LOW and used to provide the ground connection to the sensor (Figure 8-11). Note that the module is marked with a "+" on one side. Connect this side to "A1".

Figure 8-11 Sensing vibration with an Arduino

The LED is joined to a resistor as described back in Chapter 6. This can then be plugged into sockets 8 and GND on the Arduino, with the positive connection of the LED connected to 8.

Software

The software that follows uses the technique of calibrating itself as it starts, to get the "no vibration" reading from the sensor. It then waits until the sensor reading exceeds the threshold set, at which point it lights the LED. Pressing the Arduino "reset" button will cause the sensor to detect movement again.

```
// vibration_sensor

int gndPin = A0;
int sensePin = 1;
int ledPin = 8;
```

After defining the pins to use, we then define two variables. The variable "normalReading" is used during calibration (more on that in a minute), and the variable "threshold" should be set to the amount that the analog reading is allowed to exceed "normalReading" by before the LED is turned on.

```
int normalReading = 0;
int threshold = 10;
```

The "setup" function sets the appropriate pin modes and then calls the "calibrate" function to find the reading for the sensor when there is no vibration.

```
void setup()
{
  pinMode(gndPin, OUTPUT);
  digitalWrite(gndPin, LOW);
  pinMode(ledPin, OUTPUT);
  normalReading = calibrate();
}
```

The "loop" function simply takes a reading and checks to see if it has exceeded the threshold. If it has, it turns the LED on.

```
void loop()
{
  int reading = analogRead(sensePin);
  if (reading > normalReading + threshold)
  {
    digitalWrite(ledPin, HIGH);
  }
}
```

To calibrate the sensor, 100 readings are made with a one-millisecond delay between each reading, and the average is returned. A variable of type "long" is used to hold the total, as this number may be too big to fit in the usual "int" type.

```
int calibrate()
{
  int n = 100;
  long total = 0;
  for (int i = 0; i < n; i++)
  {
    total = total + analogRead(sensePin);
    delay(1);
  }
  return total / n;
}
```

How to Measure Temperature

A number of different sensor ICs are designed for measuring temperature. Perhaps the simplest to use is the TMP36 (Figure 8-12).

You can experiment with the sensor, just printing the temperature to the Serial Monitor, or you can combine the sensor with the relay module we made in Chapter 6.

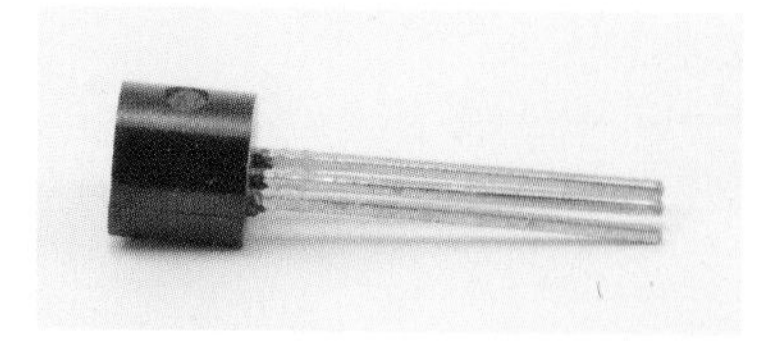

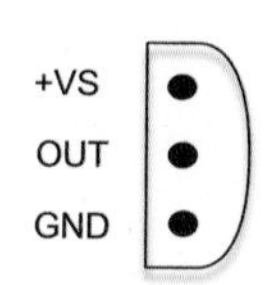

FIGURE 8-12 The TMP36

You Will Need

To use this temperature measurement IC, you will need the following items.

Quantity	Item	Appendix Code
1	Arduino Uno/Leonardo	M2/M21
1	USB lead; Type B for Uno, Micro USB for Leonardo	
1	TMP36 temperature sensor IC	S8

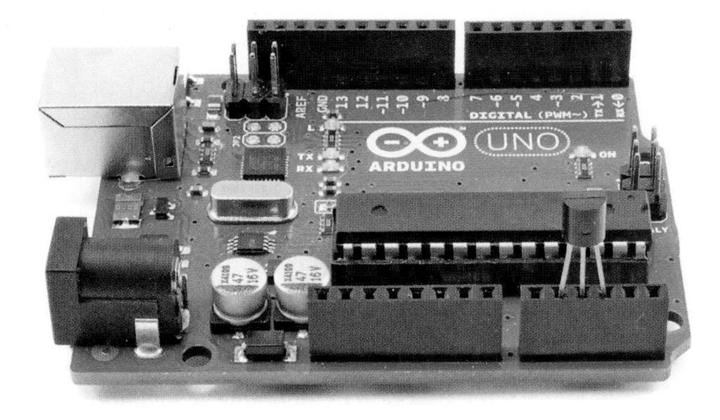

FIGURE 8-13 The TMP36 attached to an Arduino

Construction

The TMP36 has just three pins, two for the power supply and one analog output. The power supply needs to be between 2.7V and 5.5V, making it ideal for use with the 5V of an Arduino. In fact, we can supply the power to it through digital outputs and just plug the whole chip into three pins on the analog connector of the Arduino (Figure 8-13).

Software

The sketch ("temperature_sensor") follows what should now be a fairly familiar pattern. The pins are defined, and then in the "setup" function the output pins that provide power to the sensor are set to LOW for GND and HIGH for the positive supply.

```
// temperature_sensor

int gndPin = A1;
int sensePin = 2;
int plusPin = A3;

void setup()
{
  pinMode(gndPin, OUTPUT);
  digitalWrite(gndPin, LOW);
  pinMode(plusPin, OUTPUT);
  digitalWrite(plusPin, HIGH);
  Serial.begin(9600);
}
```

The main loop reads the value from the analog input and then does a bit of arithmetic to calculate the actual temperature.

First, the voltage at the analog input is calculated. This will be the raw value (between 0 and 1023) divided by 205. It is divided by 205 because a span of 1024 values occupies 5V, or 1024 / 5 = 205 per volt.

The TMP36 outputs a voltage from which the temperature in degrees C can be calculated from the equation:

tempC = 100.0 * volts – 50

For good measure, the sketch also converts this into degrees F and prints both out to the Serial Monitor.

```
void loop()
{
  int raw = analogRead(sensePin);
  float volts = raw / 205.0;
  float tempC = 100.0 * volts - 50;
  float tempF = tempC * 9.0 / 5.0 + 32.0;
  Serial.print(tempC);
  Serial.print(" C ");
  Serial.print(tempF);
  Serial.println(" F");
  delay(1000);
}
```

How to Use an Accelerometer

Tiny accelerometer modules (Figure 8-14) are now available at low cost. The two models shown are very similar, both being 5V compatible and providing analog outputs for each axis. The one on the left is from Freetronics (www.freetronics.com/am3x) and the one on the right is from Adafruit (www.adafruit.com/products/163).

FIGURE 8-14 Accelerometer modules

These modules are three axis accelerometers that measure the force applied to a tiny weight inside the chip. Two of the dimensions, X and Y are parallel to the modules PCB. The third dimension (Z) is at 90 degrees to the module's surface. There will normally be a constant force acting on this dimension due to gravity. So if you tip the module, the effect of gravity starts to increase on the dimension in which you tip it (see Figure 8-15).

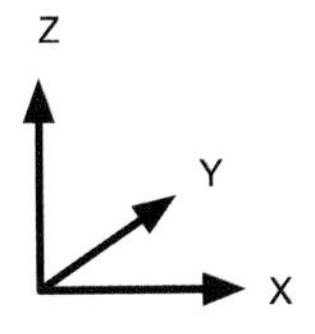

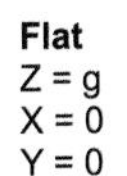

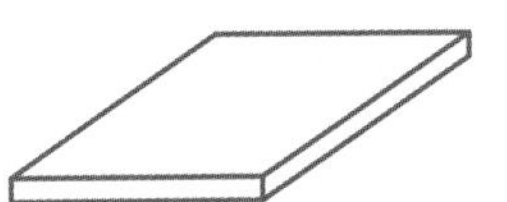

FIGURE 8-15 The effect of gravity on the accelerometer

As a vehicle to test one of these accelerometers, we are going to build an electronic version of the children's game of egg and spoon. The idea behind this is to use the accelerometer to detect the level of tilt of the "spoon" and flash an LED when it starts to be in danger of losing the egg. A buzzer sounds when the level of tilt is extreme enough for the egg to have fallen off (Figure 8-16).

You Will Need

To participate in an Arduino and spoon race, you will need the following items.

Quantity	Item	Appendix Code
1	Arduino Uno/Leonardo	M2/M21
1	USB lead; Type B for Uno, Micro USB for Leonardo	
1	Accelerometer	M15
1	Piezo buzzer	M3
1	LED	K1
1	220Ω resistor	K2
1	Battery clip to 2.1mm jack adapter	H9
1	Wooden spoon	
1	PP3 9V battery	

Construction

With a bit of thought, both of the accelerometer modules are capable of being plugged directly into the Arduino, as are the buzzer and LED. You should program the Arduino with the right sketch for the accelerometer module you are using before you attach the module, just in case some of the pins on the A0 to A5 connector are set to be outputs from a previous sketch.

FIGURE 8-16 An Arduino and spoon race

Figure 8-17 shows the schematic diagram for the Arduino Egg and Spoon.

As you can see from Figure 8-18, all the components fit into the sockets on

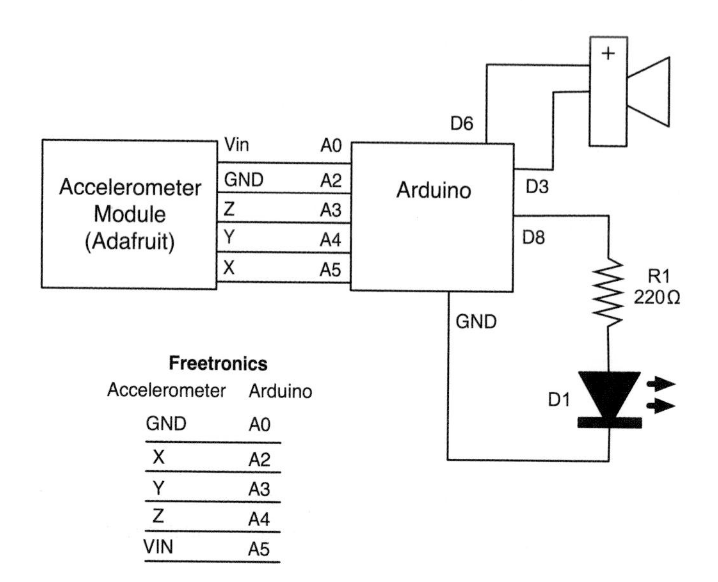

FIGURE 8-17 The schematic diagram for the Arduino Egg and Spoon

the Arduino. The LED/resistor combo is the same as we used in Chapter 6. The positive end goes to digital pin 8 on the Arduino and the negative end to GND. The buzzer fits between pins D3 and D6—D6 being connected to the positive end of the buzzer. If the pins on your buzzer are at a different spacing, then you can pick other pins, but remember to change the variables "gndPin2" and "buzzerPin" to whatever pins you end up using.

Both of the accelerometer modules will fit in the Arduino sockets A0 to A5, as shown in Figure 8-18. However, their pin allocations are quite different.

The project is powered from a 9V battery using an adapter, and the Arduino and battery are attached to the spoon with rubber bands.

FIGURE 8-18 The components attached to the Arduino

Software

There are two versions of the sketch provided: "egg_and_spoon_adafruit" and "egg_and_spoon_freetronics". Make sure you get the right one, and then program the Arduino with it BEFORE you attach the accelerometer.

The only difference between the two sketches is the pin allocations.

This is the sketch for the Adafruit version.

We start by defining the pins used.

```
// egg_and_spoon_adafruit
int gndPin1 = A2;
int gndPin2 = 3;
int xPin = 5;
int yPin = 4;
int zPin = 3;
int plusPin = A0;
int ledPin = 8;
int buzzerPin = 6;
```

The two variables "levelX" and "levelY" are used to measure the resting values of acceleration for X and Y if the spoon is level.

```
int levelX = 0;
int levelY = 0;
```

The "ledThreshold" and "buzzerThreshold" can be adjusted to set the degree of wobble before the LED lights and the buzzer sounds to indicate a "dropped egg."

```
int ledThreshold = 10;
int buzzerThreshold = 40;
```

The "setup" function initializes the pins and then calls the function "calibrate" that sets the values of "levelX" and "levelY".

```
void setup()
{
  pinMode(gndPin1, OUTPUT);
  digitalWrite(gndPin1, LOW);
  pinMode(gndPin2, OUTPUT);
  digitalWrite(gndPin2, LOW);
  pinMode(plusPin, OUTPUT);
  pinMode(ledPin, OUTPUT);
  pinMode(buzzerPin, OUTPUT);
  digitalWrite(plusPin, HIGH);
  calibrate();
}
```

In the main loop, we read the X and Y accelerations and see how much they have strayed from the values of "levelX" and "levelY". The "abs" function returns the absolute value of a number, so if the difference is negative, it is turned into a positive value, and it is this that is compared with the thresholds that have been set.

```
void loop()
{
  int x = analogRead(xPin);
  int y = analogRead(yPin);
  boolean shakey = (abs(x - levelX) > ledThreshold || abs(y - levelY) > 
ledThreshold);
  digitalWrite(ledPin, shakey);
  boolean lost = (x > levelX + buzzerThreshold || y > levelY + buzzerThreshold);
  if (lost)
  {
    tone(buzzerPin, 400);
  }
}
```

The only complication in the "calibrate" function is that we must wait for 200 milliseconds before we can take the readings. This gives the accelerometer time to turn on properly.

```
void calibrate()
{
  delay(200); // give accelerometer time to turn on
  levelX = analogRead(xPin);
  levelY = analogRead(yPin);
}
```

How to Sense Magnetic Fields

Sensing a magnetic field is made easy using a three-pin sensor IC like the A1302 linear hall effect sensor. You can use this chip in very much the same way as we did the TMP36 temperature sensor in the section "How to Measure Temperature" earlier in this chapter.

You Will Need

To use this temperature measurement IC, you will need the following items.

Quantity	Item	Appendix Code
1	Arduino Uno/Leonardo	M2/M21
1	USB lead; Type B for Uno, Micro USB for Leonardo	
1	A1302 linear hall effect sensor	S12

Construction

Just like the TMP36, the A1302 has just three pins, two for the power supply and one analog output. The power supply needs to be between 4.5V and 6V, making it ideal for use with the 5V of an Arduino.

In fact, we can supply the power to it through digital outputs and just plug the whole chip into three pins on the analog connector of the Arduino (Figure 8-19). The chip should be oriented with the dot facing outward.

Program the Arduino with the sketch before you plug in the sensor, in case A1 is set to be an output.

FIGURE 8-19 The A1302 magnetic sensor attached to an Arduino

Software

The sketch for the magnetic sensor is very similar to that of the temperature sensor.

First, the three pins are set up: digital pins 15 and 17 (A0 and A2), and A1 is set as the sensor pin.

```
// magnetic_sensor

int gndPin = A1;
int sensePin = 2;
int plusPin = A3;

void setup()
{
  pinMode(gndPin, OUTPUT);
  digitalWrite(gndPin, LOW);
  pinMode(plusPin, OUTPUT);
  digitalWrite(plusPin, HIGH);
  Serial.begin(9600);
}
```

The main loop just takes the raw reading and sends it to the Serial Monitor.

The device is not terribly sensitive, but if you hold a magnet next to it you should see a change in the reading coming from the Serial Monitor.

```
void loop()
{
  int raw = analogRead(sensePin);
  Serial.println(raw);
  delay(1000);
}
```

Summary

There are many other sensors out there, and many will interface to an Arduino quite easily using an analog input, or employing pulse length, letting you adapt the sketches used for other sensors to different sensors.

In the next chapter, we will change tack and look at sound and audio electronics.

9

Audio Hacks

In this chapter, you will look at audio electronics and find out how to make and amplify sounds so you can drive a loudspeaker.

You will also discover how to hack an FM transmitter intended for use with MP3 players in the car, so that it works as a surveillance bug.

First though, we will look at the more mundane topic of audio leads, how to use them, mend them, and make your own.

Hacking Audio Leads

Ready-to-use audio leads are pretty cheap to buy unless you go for the high-end connectors. Sometimes though, if you need a lead in a hurry, or an unusual lead, it helps to know how to wire one up from parts in your junk box or from connectors you have bought.

Many items of consumer electronics are supplied with a range of leads, and you do not always need them for use with the item you bought. Keep them in your junk box since you never know when you might need to make some kind of lead.

Figure 9-1 shows a selection of audio plugs, some designed to have leads soldered to them, and others that have plastic moldings around the lead and cable, which cannot be soldered to. Plugs with plastic moldings around them are still useful, however. It just means you will have to cut and strip the wire that leads to the plug rather than solder it to the plug itself.

General Principals

Audio leads carry audio signals, often on their way to an amplifier, and the last thing you want is for them to pick up electrical noise that will affect the quality of the sound. For this reason, audio leads are normally screened (see Figure 9-2).

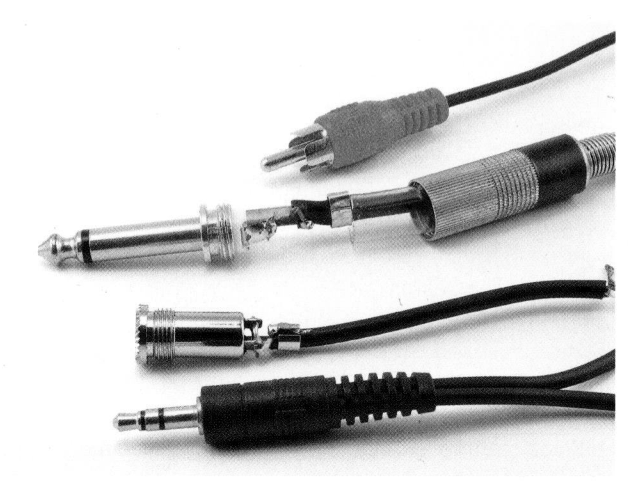

FIGURE 9-1 A variety of audio plugs

The audio signal itself (or two audio signals for stereo) is carried on insulated multi-core wires that are then enclosed in an outer conductive sheath of screening wire that carries the ground connection.

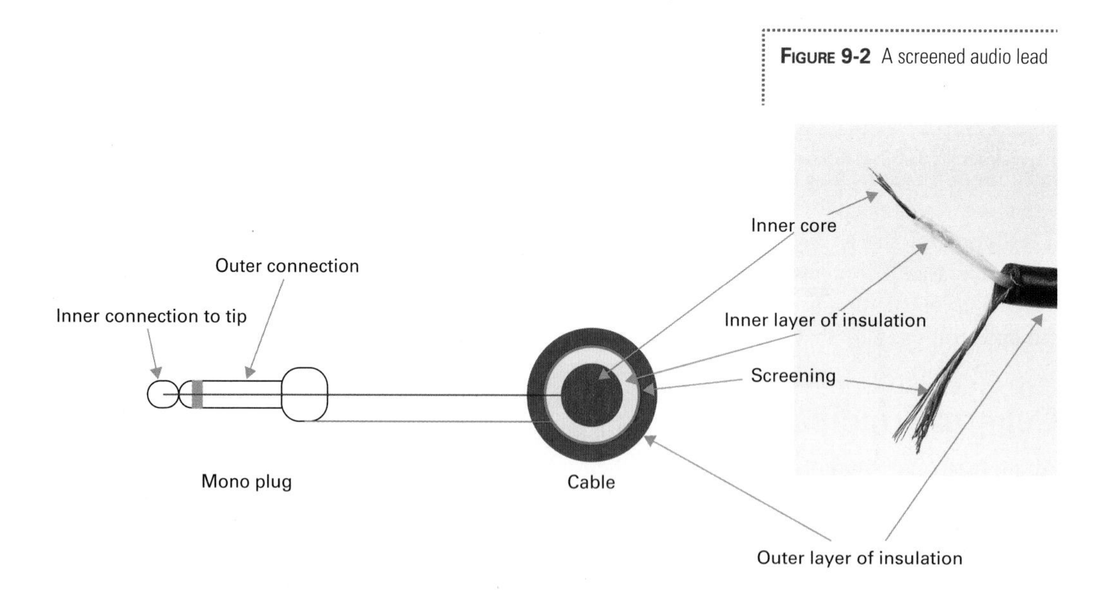

FIGURE 9-2 A screened audio lead

The exception to this is for leads to loudspeakers. These are not screened because the signal has been amplified to such a degree that any noise the speaker cables might pick up would be undetectable.

Soldering Audio Connectors

Stripping audio connectors is made more difficult by the fact that there is more than one layer of insulation. It is very easy to accidentally cut through the shielding. Nicking the outer insulation all around before stripping it will usually help with this problem.

Figure 9-3 shows the sequence involved in soldering a screened lead to a 6.3mm jack plug of the sort often used to connect an electric guitar to its amplifier.

The first step is to strip off the outer insulation about 20mm (a bit less than an inch) from the end of the lead and tease the shielding wires around to one side of the lead and twist them together. Strip about 5mm of insulation off the inner core insulation (Figure 9-3a). Then, tin both bare ends (Figure 9-3b).

The jack plug has two solder tags: one for the outer part of the plug and one connected to the tip. Both will usually have holes in them. Figure 9-3c shows the screening trimmed to a shorter length and pushed through the hole ready to solder. Once the screening is soldered into place, solder the inner core to the solder tag for the tip (Figure 9-3d).

These wires are quite delicate, so make sure the inner core wire has some extra length (as shown in Figure 9-3e) so that if the plug flexes, it will not break the connection. Notice that the strain relief tabs at the end of the plug have been pinched around the outer insulation. Finally, the plug will often have a plastic sleeve that protects the connections. Slide this over the connections and then screw in the plug casing.

Tip If there is a plug on the other end of the lead, remember to push the new plug enclosure and plastic sleeve onto the lead BEFORE you solder the second plug on, otherwise you will end up having to unsolder everything to put it on. The author has made this mistake more times than he cares to admit.

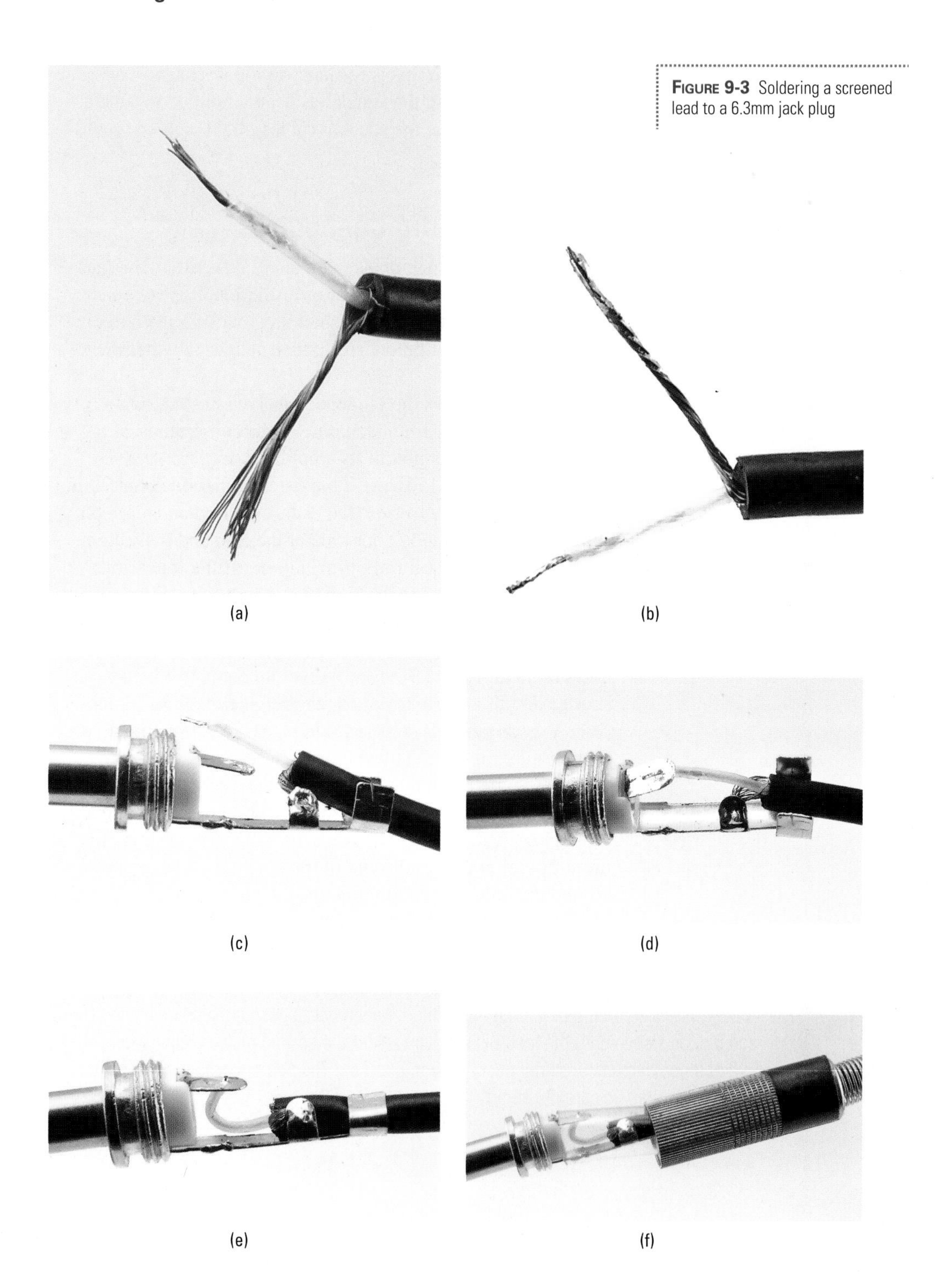

Figure 9-3 Soldering a screened lead to a 6.3mm jack plug

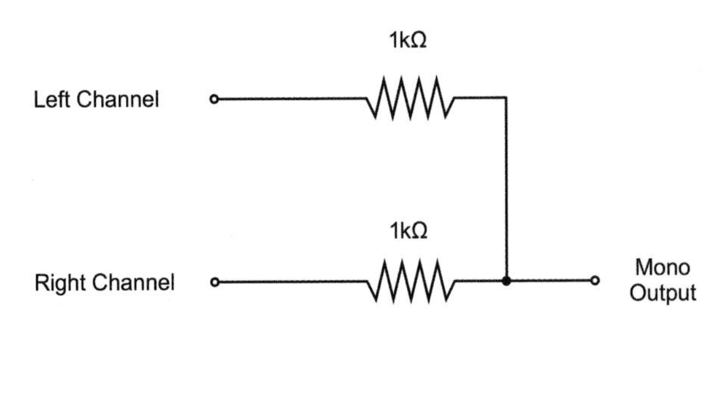

FIGURE 9-4 Mixing stereo to mono

Converting a Stereo Signal to Mono

Stereo audio is made up of two slightly different audio signals that give the stereo effect when played through two separate speakers. Sometimes, you have a stereo output that you want to input to a single channel (mono) amplifier.

You could use just one of the channels of the stereo signal (say, the left channel), but then you will lose whatever is on the right channel. So a better way of converting stereo to mono is to use a pair of resistors to mix the two channels into one (Figure 9-4).

Looking at the schematic of Figure 9-4, you could be forgiven for thinking all you need to do is connect the left and right channels to each other directly. This is not a good idea, because if the signals are very different, there is the potential for a damaging current to flow from one to the other.

As an example, we could use the mono 6.3mm jack we just soldered leads to, and combine it with a pair of resistors and a stereo 3.5mm jack plug so we could, for example, plug an MP3 player into a guitar practice amplifier.

Figure 9-5 shows the steps involved in this. To make it easier to photograph, the author's lead is made very short. You will

FIGURE 9-5 Making a lead

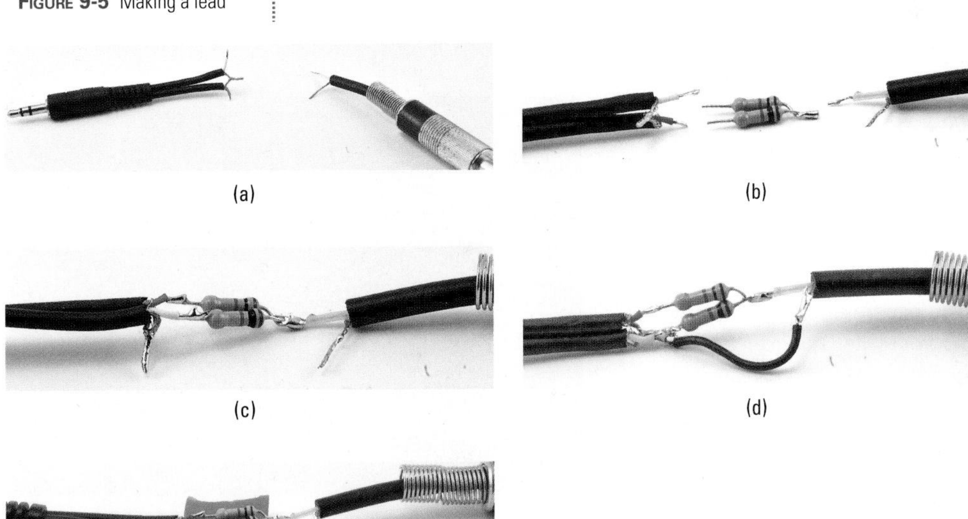

probably want to make yours longer. This is not a problem, unless you plan to make it longer than a few yards or meters.

The 3.5mm plug is of the plastic molded variety, reclaimed from some unwanted lead. The first step is to strip both leads (Figure 9-5a). Note that the stereo plug has two screened connections in one twin cable. The screened ground connections of both channels of the stereo plug can be twisted together.

Tin the ends of all the leads, and then solder the resistors together, as shown in Figure 9-5b.

Next, solder the stereo and mono leads to the resistors, as shown in Figure 9-5c, and cut and tin a short length of wire to bridge the ground connections. Solder it into place (Figure 9-5d) and then wrap everything in insulating tape, taking care to put tape in between any places where wires could short together (Figure 9-5e).

FIGURE 9-6 A microphone module

How to Use a Microphone Module

Microphones (mics) respond to sound waves, but sound waves are just small changes in air pressure, so it is not surprising that the signal you get from a mic is usually very faint. It requires amplification to bring it up to a useable level.

While it is perfectly possible to make a little amplifier to boost the signal from your mic, you can also buy a mic module that has an amplifier built in. Figure 9-6 shows such a module.

The mic module just requires a supply voltage between 2.7V and 5.5V. This makes it ideal for interfacing with an Arduino.

In Chapter 11, you will find out a bit more about oscilloscopes. But for now, here is a sneak preview of what an oscilloscope will display (see Figure 9-7) when connected to the mic module while a constant tone is being generated and the module is supplied with 5V.

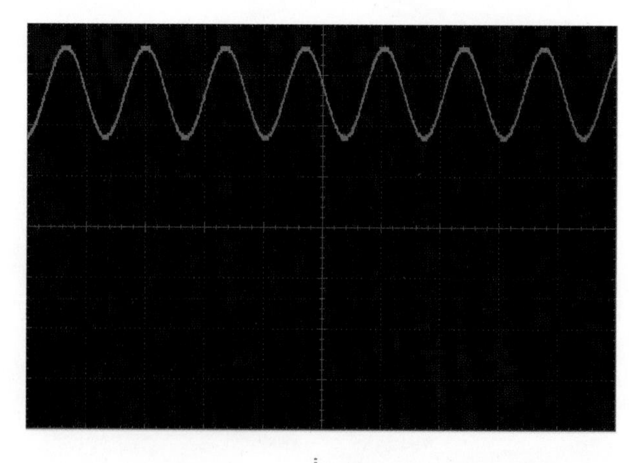

FIGURE 9-7 The output of the microphone module

The oscilloscope is displaying the sound. In this case, a constant and rather irritating tone of 7.4 kHz. The horizontal axis is time, and each blue square represents 100 microseconds. The vertical axis is the voltage and each square is 1V. The output of the mic module is a voltage that varies very quickly between about 1.8V and 3.5V. If you draw a horizontal line straight down the middle of the waveform, it would be at about 2.5V. This is halfway between 0V and 5V. So if there is no sound at all, there will just be a flat line at 2.5V, and as the sound gets louder, the waveform will swing further and further either way. It will not, however, go higher than 5V or lower than 0V. Instead, the signal will clip and become distorted.

FIGURE 9-8 The schematic diagram for a mic module

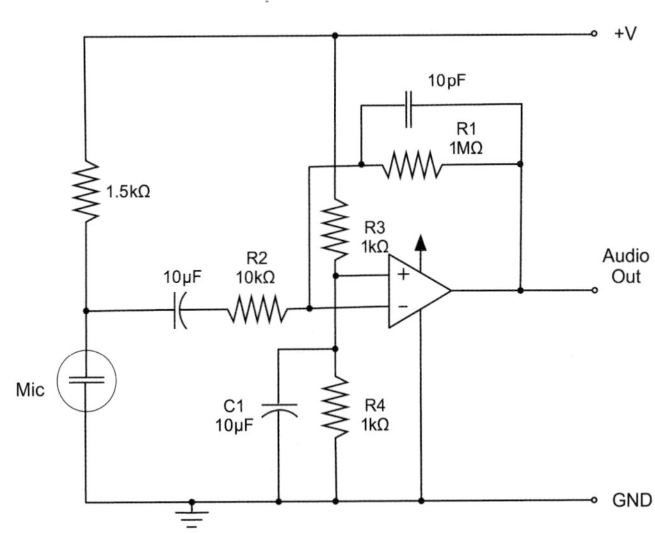

The mic module shown is sold by SparkFun (BOB-09964). The schematic for this, along with all its design files have been made public. Figure 9-8 shows a schematic for a typical microphone pre-amp.

The chip at the center of this design has a similar circuit symbol to the comparator you used in the "How to Detect Noxious Gas" section at the beginning of Chapter 8. However, it is not a comparator; it is an amplifier IC of a type known as an "operational amplifier" (or "op amp" for short).

Whereas a comparator turns its output on when the "+" input is higher than the "–" input, an op amp amplifies the difference between the "+" and "–" inputs. Left to its own devices, it amplifies this by a factor of millions. This means that the tiniest signal or noise on the input would be turned into meaningless thrashing of the output from 0 to 5V. To tame the op amp and reduce its amplification factor (called "gain"), something called "feedback" is used.

The trick is to take a portion of the output and feed it into the negative input of the op amp. This reduces the gain to an amount determined by the ratio of R1 to R2, shown in Figure 9-8. In this case, R1 is 1MΩ and R2 is 10kΩ, so the gain is 1,000,000 / 10,000 or 100.

The signal from the microphone is being amplified by a factor of 100. This shows just how weak the signal is in the first place.

The "+" input to the op amp is held halfway between GND and 5V (2.5V) by using R3 and R4 as a voltage divider. C1 helps to keep this constant.

From the schematic, you can see how you could build the module yourself on, say, stripboard. Op amps like the one used (which is a surface-mounted device) are also available in the eight-pin DIP form. However, a module like this will save you a lot of effort and may even turn out cheaper than buying and building a module from scratch.

I realize this is a rather cursory introduction to op amps. These are very useful devices, but unfortunately require more space to explain fully than this book can accommodate. You will find good information on op amps at the Wikipedia site, as well as in books with a more theoretical bent like *Practical Electronics for Inventors*, Third Edition, by Paul Sherz and Simon Monk, which has a chapter devoted exclusively to op amps.

In the next section, we will combine this module with a hacked FM transmitter of the sort used to let you play your MP3 player through your car radio, thus creating an audio "bug."

How to Make an FM Bug

To make an FM transmitter that will broadcast sound picked up from a microphone to a nearby FM radio receiver would require a lot of effort. We are hackers, so we are going to cheat and take apart an FM transmitter and wire it up to a mic module. Figure 9-9 shows the end result of this hack.

FIGURE 9-9 An FM radio bug

You Will Need

To build the bug, you will need the following.

Quantity	Item	Appendix Code
1	Microphone module	M5
1	* FM transmitter for MP3 players	
1	FM radio receiver	

* For suitable FM transmitters, try searching on eBay using the search terms "fm transmitter mp3 car." Expect to pay about USD 5 and look for the most basic of models. You do not need remote control or an SD card interface. You just want something that has an audio input lead, and for simplicity purposes runs on two AA or AAA batteries (3V).

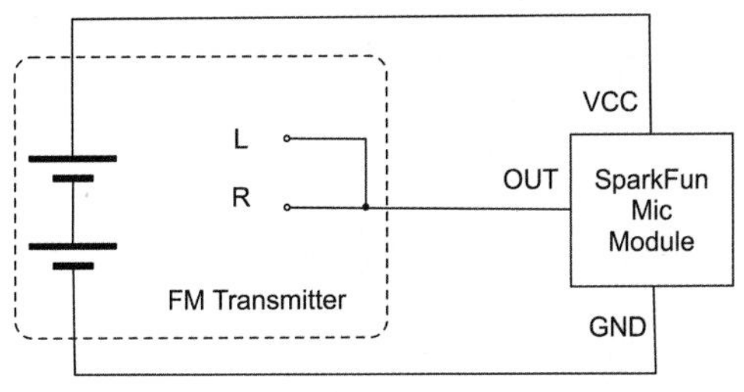

FIGURE 9-10 Schematic diagram for the radio bug

Construction

This is a very easy project to make. Figure 9-10 shows the schematic diagram for the bug.

The 3V battery of the FM transmitter is used to provide power to the mic module, and the single output of the mic module is connected to both the left and right inputs of the stereo FM transmitter.

Figure 9-11 shows how the FM transmitter is modified to connect the mic module to it.

The first step is to unscrew any screws that hold the case together and pull it apart. Then, chop off the plug, leaving most of the lead in place since the lead often doubles as an antenna in these devices. Strip and tin the three wires inside the lead (Figure 9-11a).

Looking at the three wires, in Figure 9-11a, the red wire is the right signal, the white the left, and the black ground. This is a common convention, but if you are not sure it applies to your transmitter, you can check by stripping the wires on the plug end of the lead you cut off and using the continuity setting on your multimeter to see which lead is connected to what on the plug. The farthest tip and next ring should be the left and right signals, and the metal nearest the plastic should be the ground connection.

We are going to leave the ground and left connections as they are, but disconnect the wire and connect it to the 3V connection of the battery (Figure 9-11b). In this transmitter, the positive terminal of the battery box underneath the PCB is soldered to the top surface of the PCB.

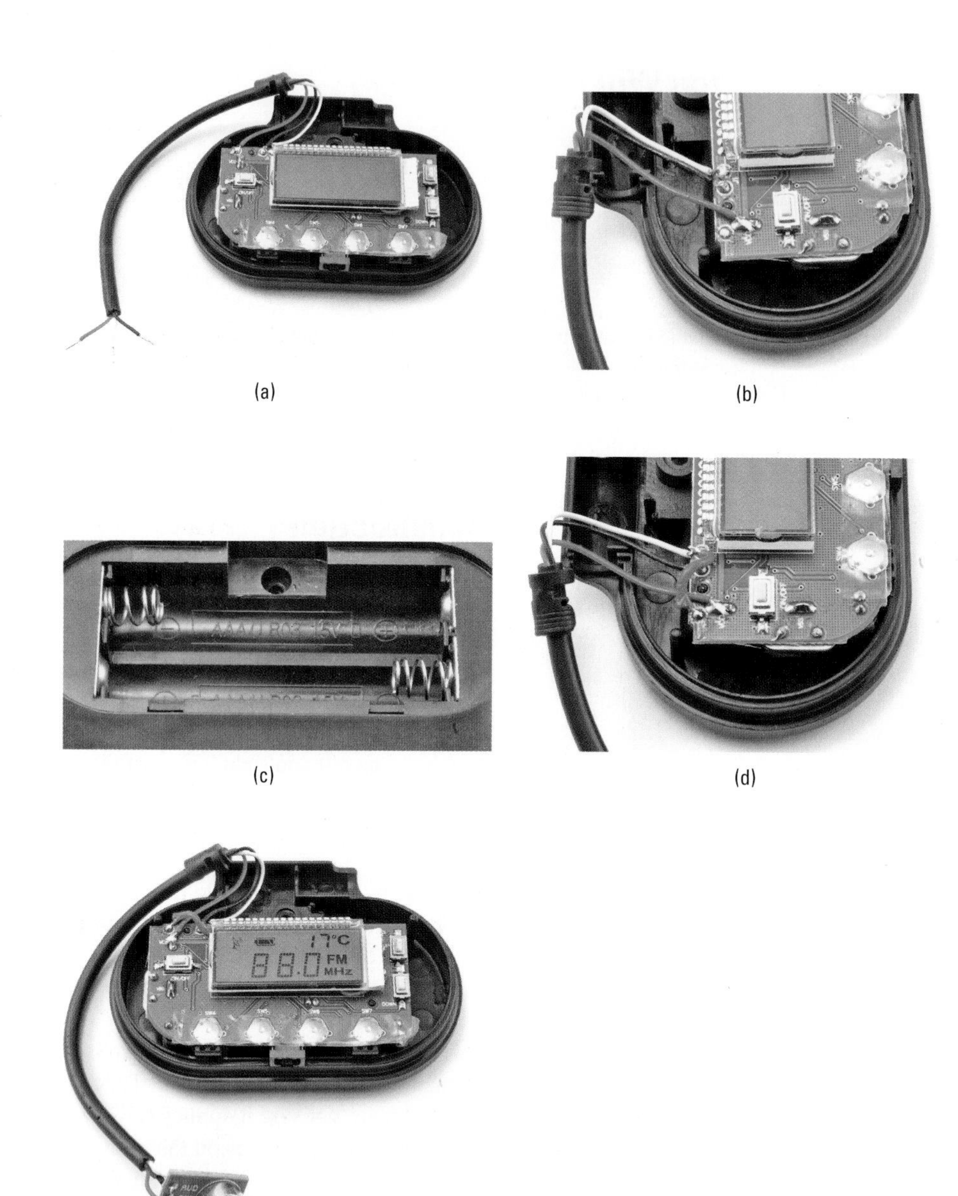

(e)

Figure 9-11 Modding the FM transmitter

To find the positive connection, look carefully at the battery box. In Figure 9-11c, you can see that the metal piece on the left of the figure links the negative of the top cell to the positive of the bottom cell. The 3V connection will therefore be the top right connection of the battery box, so trace where this comes out on the top of the PCB. If it is attached by wires, then find an appropriate place for the red wire of the audio jack lead to be joined to it.

Referring back to the schematic diagram of Figure 9-10, we need to make a little wire just to link the left and right channels (Figure 9-11d). When all the changes are complete, it should look like Figure 9-11e.

Testing

Note that the on/off button of the transmitter will have no effect on the power going to the mic module. So to fully turn off the bug, remove the batteries.

To test the module out, set the frequency of the FM transmitter to one not occupied by a radio station and then set the radio receiver to the same frequency. You may well hear the howl of feedback through the radio. To prevent that, take the radio receiver to a different room. You should find that you can hear what is happening in the room with the bug in it pretty clearly.

Selecting Loudspeakers

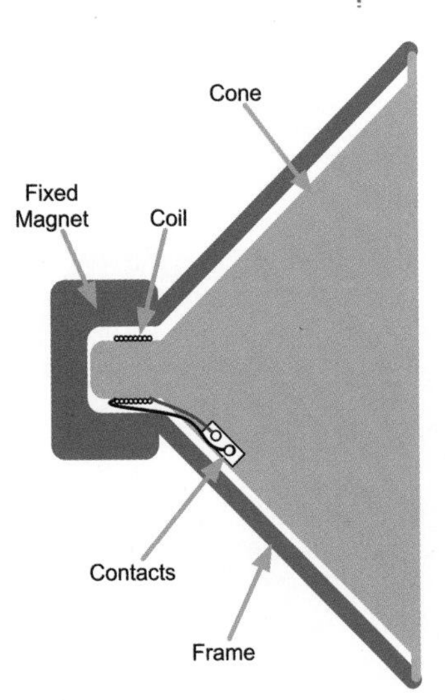

FIGURE 9-12 How a loudspeaker works

Loudspeakers have remained largely unchanged in design since the early days of radio. Figure 9-12 shows how a loudspeaker works.

The cone (often still made of paper) has a light coil around the end that sits within a fixed magnet attached to the frame of the loudspeaker. When the coil is driven by an amplified audio signal, it moves toward and away from the magnet in time with the audio. This creates pressure waves in the air, producing a sound.

Electronically speaking, a loudspeaker just looks like a coil. When you buy a speaker like this, it will have a number of ohms associated with it. Most speakers are 8Ω, but you can also commonly find 4Ω and 60Ω speakers. If you measure the resistance of the coil of an 8Ω speaker, you should find that it is indeed about 8Ω.

Another figure that is normally stated with the speaker is the power. This specifies how hard the loudspeaker can be driven before the coil will get too hot and burn out. For a small loudspeaker such as one you might put in a small radio receiver, values of 250 mW and up are not untypical. As you progress toward the kind of speakers you would use with a hi-fi set, you will see figures in the tens of watts, or even hundreds of watts.

It is very hard to build speakers that can cover the whole range of audio frequencies, which is generally standardized as 20 Hz up to 20 kHz. So you will often find hi-fi speakers that group a number of speakers into a single box. This might be a "woofer" (for low frequencies) and a "tweeter" (for high frequencies). Because woofers cannot keep up with the high frequencies, a module called a "crossover network" is used to separate the low and high frequencies and drive the two types of speakers separately. Sometimes this is taken a step further and three drive units are used: one for bass, one for mid-range tones, and a tweeter for high frequencies.

The human ear can pick out the direction of a high-frequency sound very easily. If you hear a bird tweeting in a tree, you will probably be able to look straight at it without having to think about where it is. The same is not true of low frequencies. For this reason, surround-sound systems often have a single low-frequency "woofer" and a number of other speakers that handle midrange and higher frequencies. This makes life easier, because bass speakers have to be much larger than higher-frequency units in order to push large amounts of air about relatively slowly to produce bass sounds.

How to Make a 1-Watt Audio Amplifier

Building a small amplifier to drive a loudspeaker is made easier by an IC like the TDA7052, which contains pretty much all the components you need, on a chip costing less than $1. In this section, you will make a little amplifier module on stripboard (Figure 9-13).

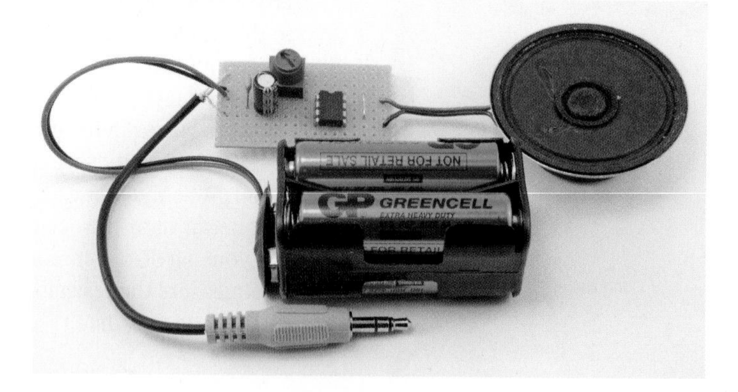

Figure 9-13 A 1-watt amplifier module

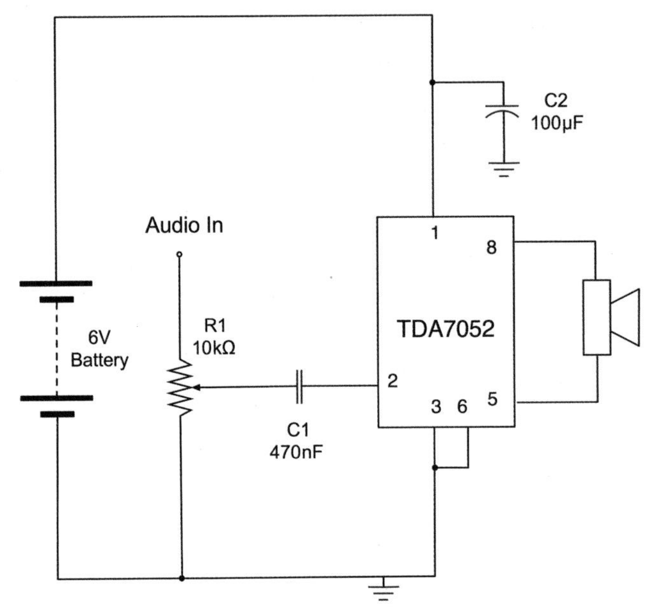

FIGURE 9-14 A typical TDA7052 amplifier schematic

An alternative to making your own amplifier is to buy a ready-made module. You will find these available for a wide range of different powers and in mono and stereo configurations. eBay is a good source for such modules, as are SparkFun (BOB-11044) and Adafruit (product ID 987). These modules often use an advanced type of design called "class-D," which is far more efficient in its use of energy than the module we are going to build.

Figure 9-14 shows the typical schematic for a TDA7052 amplifier.

R1 acts as a volume control, reducing the signal before amplification.

C1 is used to pass the audio signal on to the input to the amplifier IC without passing on any bias voltage that the signal may have from the audio device producing the signal. For this reason, when you use a capacitor like this, it is called a coupling capacitor.

C2 is used to provide a reservoir of charge that can be drawn on quickly by the amplifier when it needs it for very rapid changes in the power supplied to the speaker. This capacitor should be positioned close to the IC.

You Will Need

To build the amplifier module, you will need the following.

Quantity	Name	Item	Appendix Code
1	IC1	TDA7052	S9
1	R1	10kΩ variable resistor	K1, R1
1	C1	470nF capacitor	C3
1	C2	100µF capacitor	K1, C2
1		8Ω speaker	H14
1		Stripboard	H3

Construction

Figure 9-15 shows the stripboard layout for the amplifier module. If you have not used stripboard before, read through the section titled "How to Use Stripboard (LED Flasher)" in Chapter 4.

To build the module, follow the steps shown in Figure 9-16.

First, cut the stripboard to size and make the three cuts in the tracks using a drill bit (Figure 9-16a).

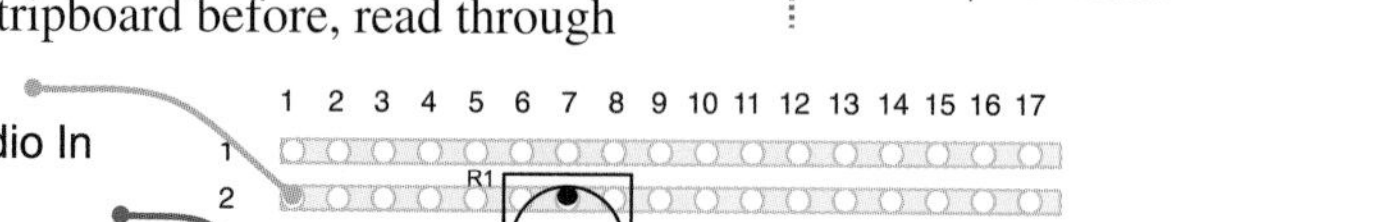
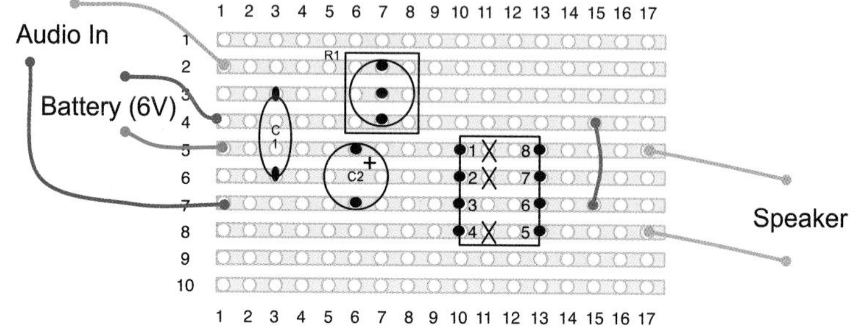

FIGURE 9-15 The stripboard layout for an amplifier module

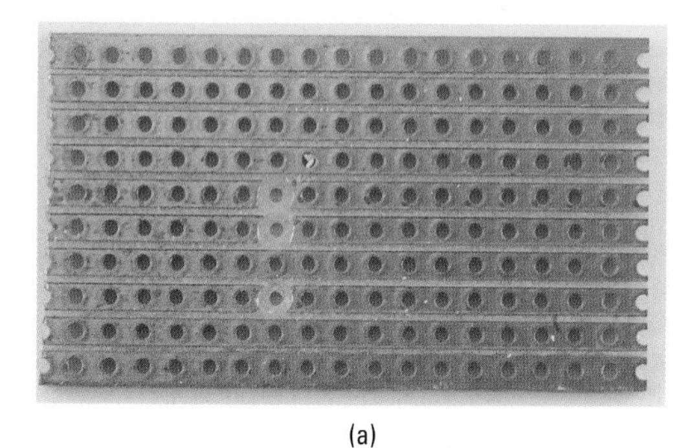

(a)

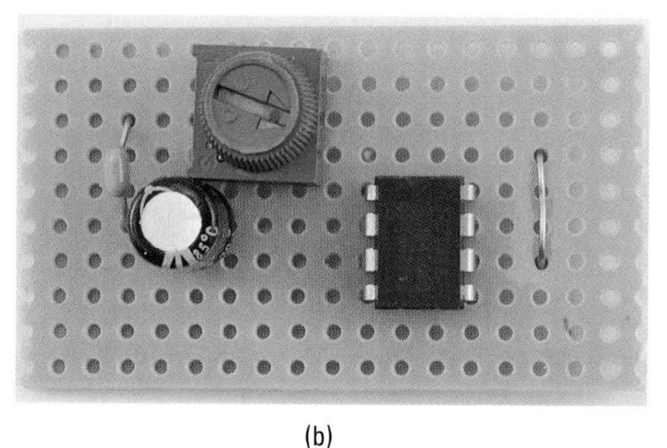

(b)

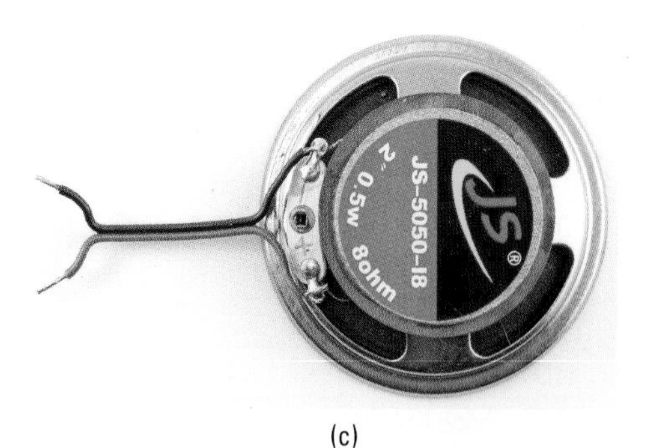

(c)

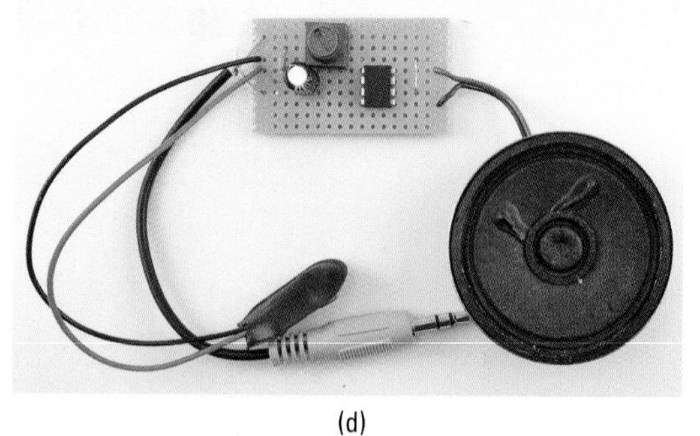

(d)

FIGURE 9-16 Building the audio amplifier module

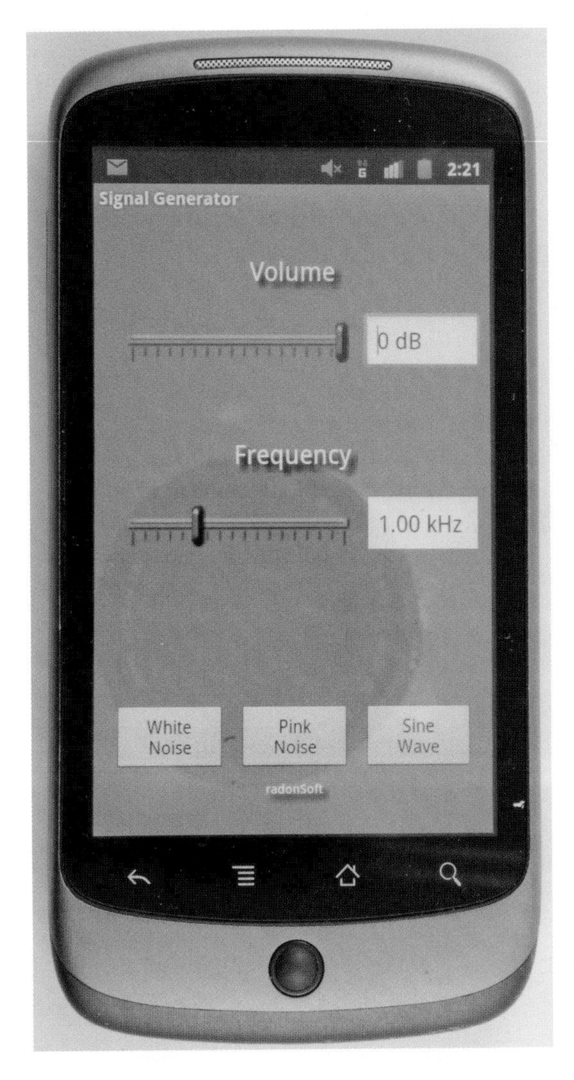

FIGURE 9-17 A signal generator app

The next step is to solder the link into place, and then the IC, C1, C2, and R1 in that order (Figure 9-16b). It is easiest to solder the components that are lowest to the board first.

Attach leads to the speaker (Figure 9-16c) and finally attach the battery clip and a lead ending in a 3.5mm stereo jack plug (Figure 9-16d). Note that only one channel of the audio lead is used. If you want to use both left and right channels, you should use a pair of resistors (see the section "Hacking Audio Leads" at the beginning of this chapter).

Testing

You can try the amplifier out by plugging it into an MP3 player, or, if you have an Android phone or an iPhone, download a signal generator app like the one shown in Figure 9-17. There are a number of such apps, many of them free, including this one for Android from RadonSoft.

With this, you can play a tone at a frequency you select. By noting when the volume of the speaker starts to drop off, you can work out the useful frequency range of your amplifier module.

How to Generate Tones with a 555 Timer

Back in Chapter 4, you used a 555 timer to blink a pair of LEDs. In this section, we will see how to use a 555 timer IC oscillating at much higher frequencies to generate audio tones.

The pitch will be controlled using a light-dependent resistor (LDR) so that as you wave your hand over the light sensor, the pitch will change in a theremin-like manner.

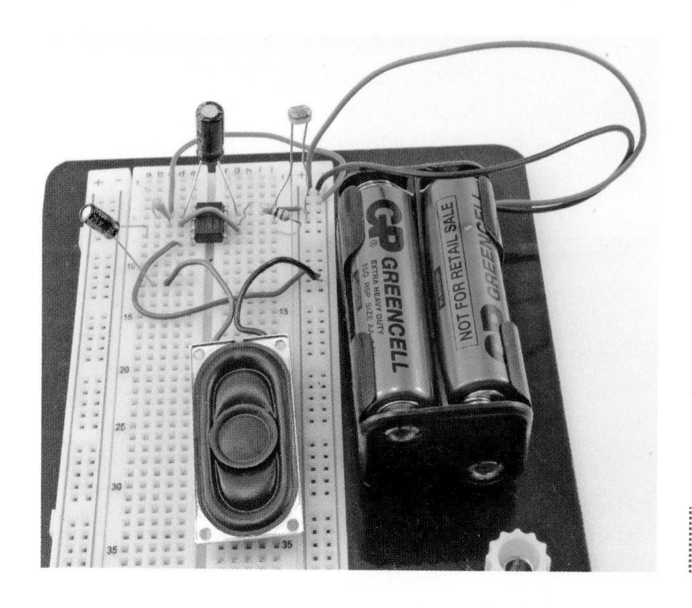

Figure 9-18 Generating tones with a 555 timer IC

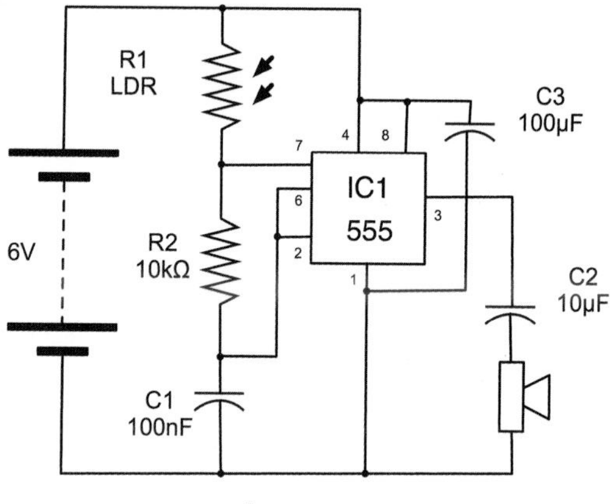

Figure 9-18 shows the tone generator built onto breadboard.

Figure 9-19 shows the schematic diagram for the tone generator.

This is similar to the design of the LED flasher in Chapter 4. In this case, instead of two fixed resistors and a capacitor setting the frequency, R1 is the LDR, whose resistance will vary between about 1kΩ and 4kΩ depending on the light falling on it. We need a much higher frequency than our LED flashing circuit—in fact, if we aim for a maximum frequency of around 1 kHz, we need a frequency of about 1000 times what we had before.

Figure 9-19 Schematic diagram for a 555 tone generator

The 555 timer oscillates at a frequency determined by the formula:

frequency = 1.44 / ((R1 + 2 * R2) * C)

where the units of R1, R2, and C1 are in Ω and F.

So, if we use a 100nF capacitor for C1, and R2 is 10kΩ, and R1 (the LDR) has a minimum frequency of 1kΩ, then we can expect a frequency of:

1.44 / ((1000 + 20000) * 0.0000001) = 686 Hz

If the LDR's resistance increases to 4kΩ, then the frequency will drop to:

1.44 / ((4000 + 20000) * 0.0000001) = 320 Hz

To calculate the frequency and when deciding what values of R1, R2, and C1 to use, there are online calculators like this one at www.bowdenshobbycircuits.info/555.htm that will calculate the frequency for you.

You Will Need

To build the amplifier module, you will need the following.

Quantity	Name	Item	Appendix Code
1	IC1	555 timer IC	K1, S10
1	R1	LDR	K1, R2
1	R2	10kΩ resistor	K2
1	C1	100nF capacitor	K1, C4
1	C2	10μF capacitor	K1, C5
1		8Ω speaker	H14

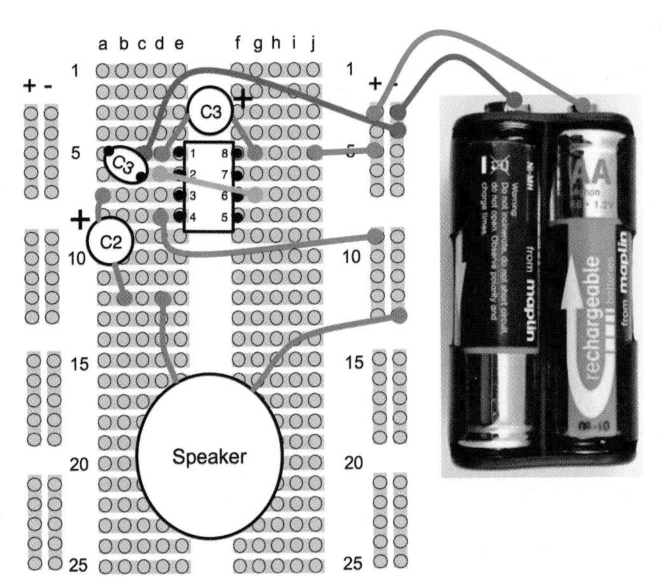

FIGURE 9-20 A signal generator app

Construction

Figure 9-20 shows the breadboard layout for the tone generator.

It would be quite straightforward to build this design onto stripboard. The stripboard layout in the section "How to Use Stripboard (LED Flasher)" in Chapter 4 would be a good starting point.

How to Make a USB Music Controller

Music software like Ableton Live™ is designed to allow USB controllers that emulate a keyboard to control virtual musical instruments and do all kinds of exciting things.

You can use the USB keyboard emulation features of the Arduino Leonardo with an accelerometer so that tilting the board produces a key press of a number between 0 and 8, with 4 being pressed if the board is level, 0 if tilted almost vertically to the right, and 8 being pressed when it is tilted the other way.

The only hardware on the Arduino is the accelerometer (Figure 9-21).

Figure 9-21 A USB music controller

You Will Need

To build this controller, you will need the following items.

Quantity	Item	Appendix Code
1	Arduino Leonardo	M21
1	Micro USB for Leonardo	
1	Accelerometer	M15 (Adafruit version)

Construction

There is actually very little to construct in this project. The schematic is actually the same as in the section titled "How to Use an Accelerometer" in Chapter 8. The Freetronics accelerator will also work, but you will need to change the pin assignments before attaching the accelerometer.

Software

The software for the music controller combines code for sensing the angle of tilt on the X-axis with emulating a keyboard press.

The first step is to assign the pins to be used. As in the section "How to Use an Accelerometer" in Chapter 8, the accelerometer module is powered from output pins.

```
// music_controller

int gndPin = A2;
int xPin = 5;
```

```
int yPin = 4;
int zPin = 3;
int plusPin = A0;
```

The variable “levelX” is used during calibration and holds the analog value when the accelerometer is flat.

The “oldTilt” variable contains the old value of the tilt of the board, which is a value between 0 and 8, where 4 means level. The old value is remembered, so that a key press is only sent if the tilt angle changes.

```
int levelX = 0;
int oldTilt = 4;
```

The “setup” function sets the output pins to power the accelerometer, calls “calibrate”, and starts the Leonardo keyboard emulation mode.

```
void setup()
{
  pinMode(gndPin, OUTPUT);
  digitalWrite(gndPin, LOW);

  pinMode(plusPin, OUTPUT);
  digitalWrite(plusPin, HIGH);
  calibrate();
  Keyboard.begin();
}
```

In the main loop, the accelerometer reading is converted to a number between 0 and 8, and if it has changed since the last reading, a key press is generated.

```
void loop()
{
  int x = analogRead(xPin);
  // levelX-70 levelX levelX + 70
  int tilt = (x - levelX) / 14 + 4;
  if (tilt < 0) tilt = 0;
  if (tilt > 8) tilt = 8;
  // 0 left, 4 is level, 8 is right
  if (tilt != oldTilt)
  {
      Keyboard.print(tilt);
      oldTilt = tilt;
  }
}
```

The “calibrate” function takes an initial reading of the acceleration on the X-axis, after waiting 200 milliseconds for the accelerometer to turn on properly.

```
void calibrate()
{
  delay(200); // give accelerometer time to turn on
  levelX = analogRead(xPin);
}
```

How to Make a Software VU Meter

The mic module you used in the section “How to Make an FM Bug” is also perfectly suited for use with microcontrollers like the Arduino. Figure 9-22 shows the module with pins attached to it and pushed into the analog connector strip of the Arduino.

The mic module can be used to measure the sound level and write a number of “*”s to the Serial Monitor to indicate the loudness of the sound (Figure 9-23).

FIGURE 9-22 Attaching a mic module to an Arduino

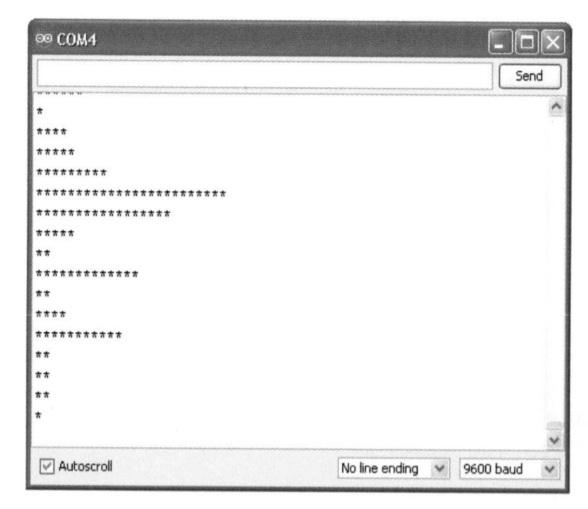

FIGURE 9-23 The Serial Monitor as a VU meter

You Will Need

To build this VU meter, you will need the following items.

Quantity	Item	Appendix Code
1	Arduino Uno/Leonardo	M2, M21
1	USB lead; Type B for Uno, Micro USB for Leonardo	
1	Mic module	M14
1	Header pins (three-way)	H4

Construction

Upload the sketch "vu_meter" before attaching the mic module.

Solder the header pins to the module so they will fit into the Arduino sockets A0 to A2 with the microphone facing outward, as shown in Figure 9-22.

Software

The mic module uses very little current, so for convenience we can use A0 and A1 to provide the power to it.

The sketch begins by defining the pins to use and setting them up in the "setup" function. Serial communication is also started here.

```
// vu_meter

int gndPin = A1;
int plusPin = A0;
int soundPin = 2;

void setup()
{
  pinMode(gndPin, OUTPUT);
  digitalWrite(gndPin, LOW);
  pinMode(plusPin, OUTPUT);
  digitalWrite(plusPin, HIGH);
  Serial.begin(9600);
}
```

The loop function reads the raw value for the analog input A2. The mic module produces an output of 2.5V when there is no signal, and swings above and blows that with the sound's waveform. So to find the "loudness" we need to first subtract 511 from the raw value—511 being equivalent to the 2.5V offset in the raw analog reading that spans from 0 to 1023.

The "abs" function makes any negative number positive and then divides the whole result by 10 to give us a number between 0 and 51 and assigns it to the variable "topLED". We are not actually using LEDs, but you could think of each "*" as being an LED illuminated on a bar graph display.

The "for" loop then prints a number of "*"s equal to the value held in "topLED". Finally, a new line is printed and we delay for 1/10th of a second.

```
void loop()
{
  int value = analogRead(soundPin);
  int topLED = 1 + abs(value - 511) / 10;

  for (int i = 0; i < topLED; i++)
  {
      Serial.print("*");
  }
  Serial.println();
  delay(100);
}
```

Summary

In addition to the how-tos just covered, there are lots of audio modules you can make use of. Low-cost stereo power amplifiers are available from eBay and suppliers like SparkFun and Adafruit.

You can also buy ultra-low-cost amplified speakers intended for computers and reuse them in your projects.

10

Mending and Breaking Electronics

In this chapter, we will look at taking things apart and putting them back together again, or just taking them apart to salvage components.

In today's throw-away society, many consumer electronics items that stop working go directly into the garbage. Economically, they are simply not worth paying someone to repair. However, that does not mean it is not worth *trying* to repair them. Even if the attempt fails, some serviceable components may be scavenged for use in your projects.

How to Avoid Electrocution

When working on something that is powered by household electricity, NEVER work on it when it is plugged into the outlet. I actually like to have the electrical plug for the appliance right in front of me, so that I know it is not plugged in. Household electricity kills many people every year. Take it seriously!

Some devices, such as switch mode power supplies, contain high-value capacitors that will hold their charge for hours after the device has been unplugged. These capacitors are simply biding their time, waiting for some unsuspecting fingers to complete the circuit.

Unless it is a very small capacitor, it should not be discharged by shorting the leads with a screwdriver. A large capacitor at high voltage can supply huge amounts of charge in a fraction of a second, melting the end of the screwdriver and flinging molten metal around. People have been blinded by capacitors exploding in this manner, so don't do it.

Figure 10-1 shows the safe way to discharge a capacitor.

FIGURE 10-1 Safely discharging a capacitor

The legs of a 100Ω resistor are bent to the right fit for the capacitor contacts and held in the teeth of a pair of pliers for a few seconds. You can use the highest setting of your voltmeter to check that the capacitor has discharged to a safe level (say, 50V). If you have a high-wattage resistor, all the better. If it is not high enough power, it will break, but not in as spectacular a way as a capacitor being discharged dangerously.

Some devices that can pack a painful and sometimes lethal punch are:

- Old glass CRT TVs
- Switch-mode power supplies
- Camera flash guns and disposable cameras with a flash

How to Take Something Apart AND Put It Back Together Again

It is often said that "any fool can take something apart, but putting it back together is a totally different matter."

Just remember that taking things apart usually voids their warranty.

By following a few simple rules, you shouldn't have any problems.

- Have a clear working area with lots of room.
- As you take out the screws, place them in the same pattern on your worktop as they were in the case they came out of. Sometimes the screws can be different sizes. If they are likely to be knocked or roll about on the surface, then push them into a piece of expanded polystyrene or something similar.
- After undoing the screws, when you come to take the case apart, watch out for any little plastic bits like switch buttons that might fall out. Try and keep them in place until you are ready to remove them.
- If something looks tricky, draw a sketch or take a photograph. (I tend to take a lot of photographs when repairing things, like with a hair dryer or straighteners, that have a large mechanical design component.)
- Try not to force things apart. Look to see where the clips are.
- If all else fails, try cutting the case apart with a handsaw (something your author has resorted to in the past), and then later glue the case back together.

How to Check a Fuse

The most convenient problem to fix in an appliance is the fuse. It's convenient because it is easy to test and easy to fix. Fuses are basically just wires designed to burn out when the current flowing through them gets too high. This prevents further damage to more expensive components, or can stop a fire from starting.

Sometimes fuses are clear, so you can see that the wire inside them has broken and that they have "blown." Fuses are rated in amps and will generally be labeled to show the maximum current in A or mA they can take. Fuses also come as "fast blow" and "slow blow." As you would expect, this determines how fast the fuses react to over-current.

Some household electrical plugs contain a fuse holder, and you can also find fuses on PCBs. Figures 10-2a-c show the

(a) (b)

(c)

Figure 10-2 Fuses

inside of a UK fused plug and also a fuse holder on the PCB for the author's multimeter.

You have used your multimeter in Continuity mode enough times now that you can probably guess how to test a fuse (Figure 10-3).

If a fuse has blown, there may be a good reason for this. Occasionally, however, they blow for other reasons, such as a momentary spike in the electric power lines or when turning on a heating element on a particularly cold day. So, generally, if there is no obvious sign of a problem with the device

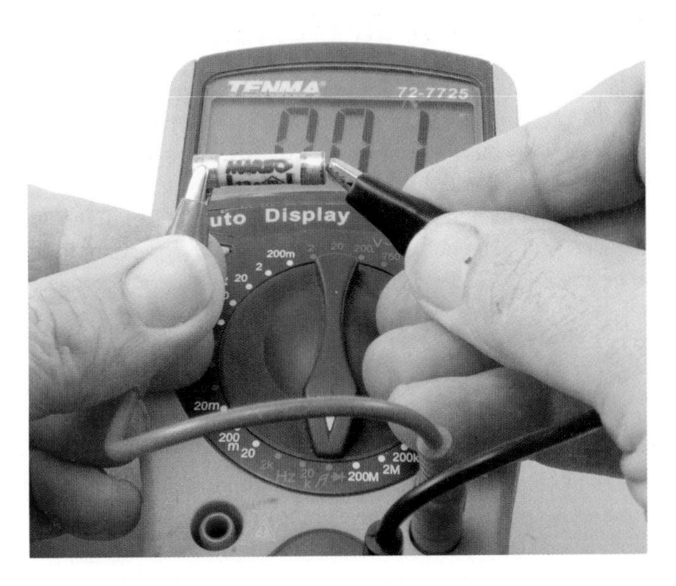

FIGURE 10-3 Testing a fuse with a multimeter

(look for wires that have come lose, or any sign of charring), then try replacing the fuse.

If the fuse immediately blows again, don't try another. You should instead find the source of the problem.

How to Test a Battery

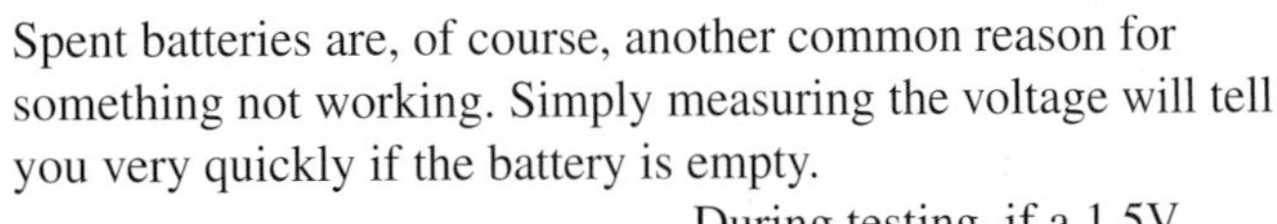

Spent batteries are, of course, another common reason for something not working. Simply measuring the voltage will tell you very quickly if the battery is empty.

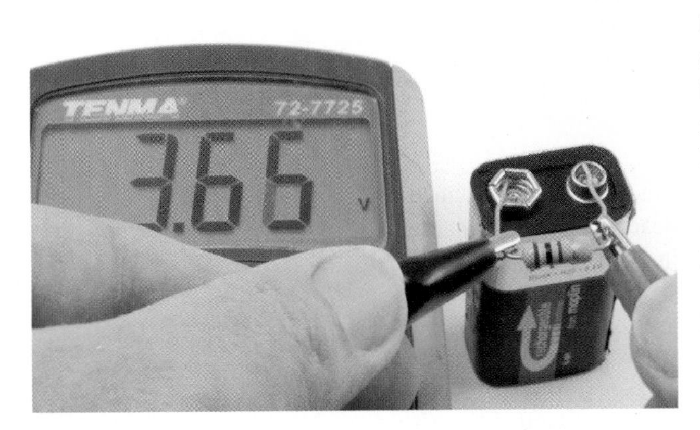

FIGURE 10-4 Testing a battery using a resistor and multimeter

During testing, if a 1.5V battery like an AA or AAA is showing less than 1.2V, or a 9V battery is showing less than 8V, it is probably time to throw it away. However, the voltage of a battery shown when it is not powering anything can be a little misleading. For a more accurate picture, use a 100Ω resistor as a "dummy" load. Figure 10-4 shows a resistor and multimeter being used to assess the state of the battery.

How to Test a Heating Element

If you have a suspect heating element from an oven, hair dryer, or so on, you can check it by measuring its resistance. As with anything using household electricity, only do this when the appliance is completely disconnected.

It's a good idea to roughly work out what you think the resistance should be before you measure it. So, for example, if you have a 2-kW 220V heating element, then rearrange:

$$P = V^2 / R$$

to

$$R = V^2 / P = 220 \times 220 / 2000 = 24\Omega$$

Calculating what you expect before you measure it is always a good idea, because if you measure it first, it is all too easy to convince yourself that it was what you were expecting. For instance, one time your humble author convinced himself that a suspect element was fine because it was showing a resistance of a few hundred ohms. Eventually, it transpired that there was a light bulb in parallel with the heating element and that the element itself was instead broken.

Finding and Replacing Failed Components

When something stops working on a PCB, it is often the result of something burning out. This sometimes leads to charring around the component. Resistors and transistors are common culprits.

Testing Components

Resistors are easy to test with a multimeter set to its resistance range. Although the results can be misleading, you can test them without removing them. Most of the time, you are looking for an open circuit, very high resistance, or sometimes a short (0Ω).

If your multimeter has a capacitance range, these too can easily be tested.

Other components are less easily identified. It is usually possible to make out some kind of device name on the case.

A magnifying glass is sometimes useful, as is taking a digital photograph and then zooming in to a high magnification. Having found some kind of identifying mark, type it into your favorite search engine.

Bipolar transistors can also be tested (see the section "How to Use a Multimeter to Test a Transistor" in Chapter 11). However, if you have a spare, it is often easier just to replace it.

Desoldering

There is definitely a knack to desoldering. You often have to add more solder to get the solder to flow. I find it quite effective to draw the solder off onto the tip of the soldering iron, which I keep cleaning using the sponge.

Desoldering braid (Appendix – code T13) is also quite effective. Figure 10-5 shows the steps involved in using desoldering braid to remove the solder from around a component lead so it can be removed.

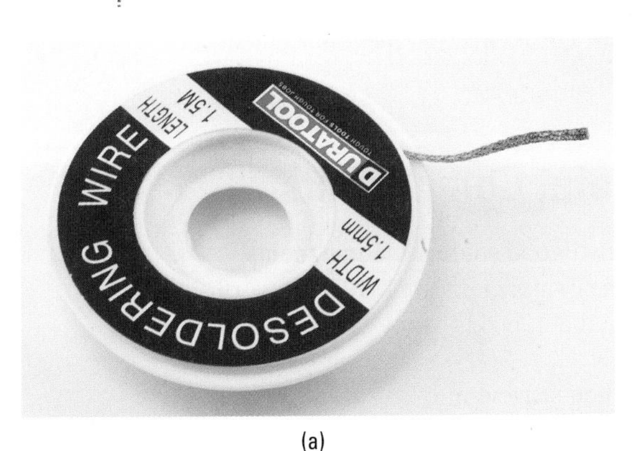

(a)

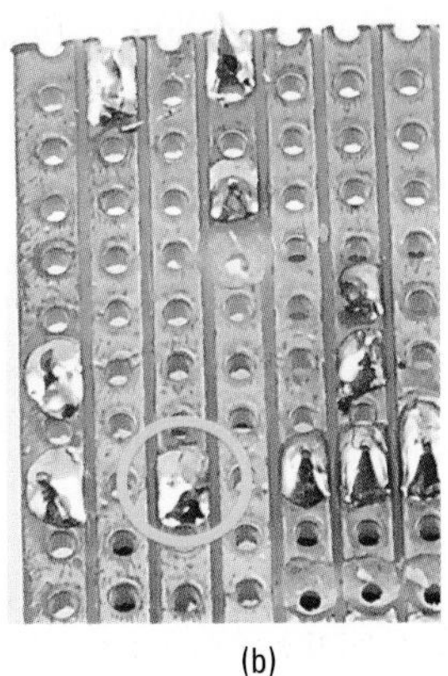

(b)

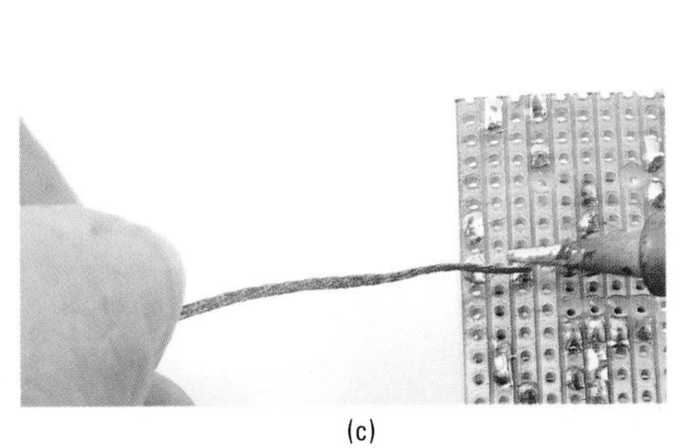

(c)

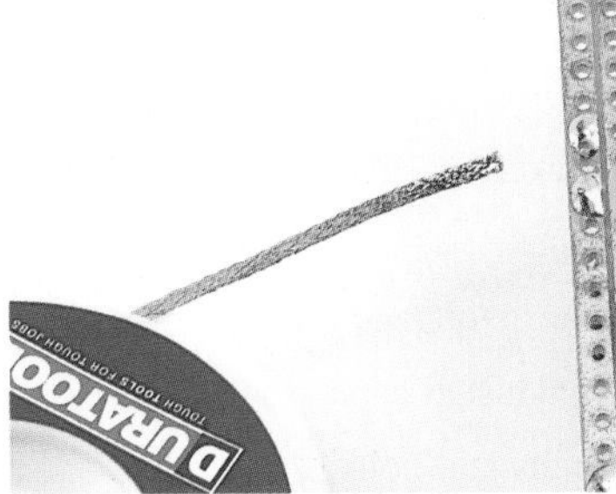

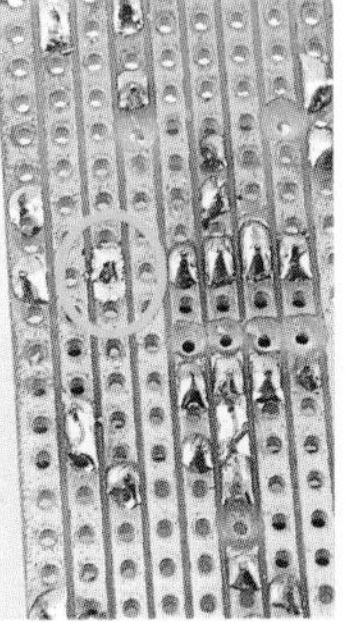

(d)

FIGURE 10-5 Using desoldering braid

Desoldering braid (Figure 10-5a) is supplied in small lengths on a small reel. You do not need much. It is braided wire impregnated with flux that encourages the solder to flow into it and off the PCB or stripboard copper.

Figure 10-5b shows the joint (circled in yellow) that we are going to remove the solder from. Press the braid onto the joint with the soldering iron (Figure 10-5c) and you should feel the blob of solder on the joint start to melt into the braid. Remove the braid while everything is hot and you should see a nice clean joint with the solder transferred to the braid (Figure 10-5d).

Cut off the section of the braid with solder on it and throw it away.

You may have to do this a couple of times to remove enough solder to release the component.

Replacement

Soldering in the replacement component is straightforward, you just have to make sure you get it the right way around. This is where photographing the board before making the replacement can be a good idea.

How to Scavenge Useful Components

Dead consumer electronics are a good source of components. But be selective, because some components are really not worth saving. Resistors are so cheap that it is really not worth the effort of removing them.

Here is what I look for when scavenging:

- Any kind of motors
- Connectors
- Hookup wire
- Seven-segment LED displays
- Loudspeakers
- Switches
- Large transistors and diodes
- Large or unusual capacitors
- Screws nuts and bolts

Figure 10-6 shows the insides of a dead video cassette recorder, with some of the more interesting parts for scavenging labeled.

FIGURE 10-6 Scavenging from a VCR

The easiest way to remove a lot of components, and things like hookup wire, is simply to snip them with wire cutters. The same applies to large electrolytic capacitors and other items, as long as they still leave leads long enough to use. Alternatively, you can desolder the items.

How to Reuse a Cell Phone Power Adapter

Everything you make in electronics requires a power source of some sort. Sometimes this will be batteries, but often it is more convenient to power the device from your household electricity.

Given that most of us have drawers stuffed with obsolete mobile phones and their charges, it makes sense to be able to reuse an old mobile phone charger. If they are newish phones, they may well have some kind of standard connector on them like a mini-USB or micro-USB, but many older phone models had a proprietary plug, used only by that phone manufacturer.

There is nothing to stop us from taking such an adapter and putting a more standard plug on the end of it, or even connecting the bare wires to screw terminals.

Figure 10-7 shows the steps involved in putting a different type of connector, such as a 2.1mm barrel jack, on the end of an old cell phone charger.

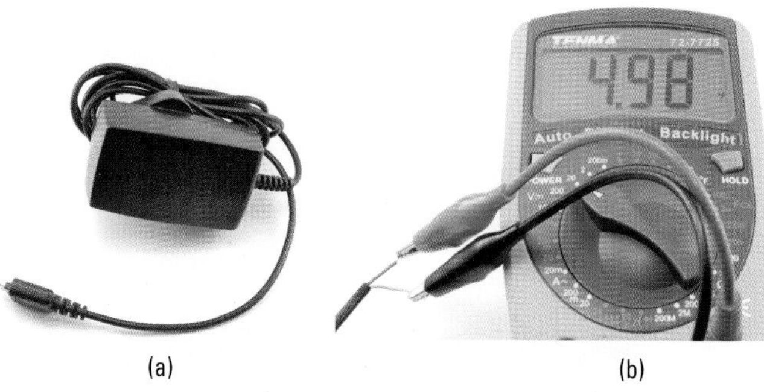

The charger is of the "wall-wart" type that plugs directly into an electrical outlet. The connector is of a type long since discontinued (Figure 10-7a). The charger has a label saying that it can supply 5V at 700mA, so the first step is (making sure the charger is unplugged) to chop off the existing connector and strip the bare wires. There should be two wires, and if one is black and one is red, then the red one is usually positive and the black negative. In this case, the wires are red and yellow. Whatever colors the wires are, it is always a good idea to use a multimeter to check the polarity (Figure 10-7b).

(c) (d)

FIGURE 10-7 Attaching a barrel jack to a cell phone charger

Remember to put the lead through the plastic body of the barrel jack before you start soldering!

You can then solder on a barrel jack plug (Appendix–code H11). This is much the same procedure we used for an audio lead in the section "Hacking Audio Leads" in Chapter 9. Figure 10-7c shows the plug ready to solder, while Figure 10-7d displays the final lead ready to use.

Summary

In this chapter, we have discovered some of the treasures that can be rescued from dead electronic equipment and also briefly looked at testing and mending.

If you want to learn more about mending things, I recommend the book *How to Diagnose and Fix Everything Electronic* by Michael Geier (McGraw-Hill/TAB, 2011).

11

Tools

This chapter is mainly for reference. You have already met some of the techniques described here while working your way through the book.

How to Use a Multimeter (General)

Figure 11-1 shows a close-up of the range selector of my multimeter.

This is typical of a medium-range multimeter costing around USD 20. We have probably only used four or five of the settings during the course of this book, so it is worth pointing out some of the other features of a multimeter like this.

Continuity and Diode Test

Starting at 6 o'clock, we have the Continuity mode, represented by a little music symbol and also a diode symbol. We have used the Continuity mode many times. It just beeps when there is very low resistance between the leads.

The reason a diode symbol appears here is because this mode also doubles for testing diodes. With some multimeters, this feature will also work on LEDs, allowing you to measure the forward voltage.

Connect the anode of the diode (the end without a stripe in a normal diode, and with a longer lead on an LED) to the red test lead of the multimeter, and then the other end of the diode to the black lead. The meter will then tell you the forward voltage of the diode. So, expect to see about 0.5V for a normal diode and 1.7V to 2.5V for an LED. You will probably also find that the LED glows a little.

Figure 11-1 Multimeter range selection

Resistance

The multimeter in Figure 11-1 has six resistance ranges, from 200MΩ down to 200Ω. If you pick a range that has a maximum resistance lower than the resistor you are measuring, then the meter will indicate this. Mine does so by displaying a "1" on its own without any further digits. This tells me I need to switch to a higher resistance range. Even better, start at the maximum range and work your way down until you get a precise reading. For the most precise reading, you need the meter to be on the range above the one that tells you it's out of range.

When measuring high-value resistors of 100kΩ and up, remember that you yourself are also a big resistor, so if you hold the test lead to the resistor at both ends (see Figure 11-2), you are measuring both the resistor in question and your own resistance.

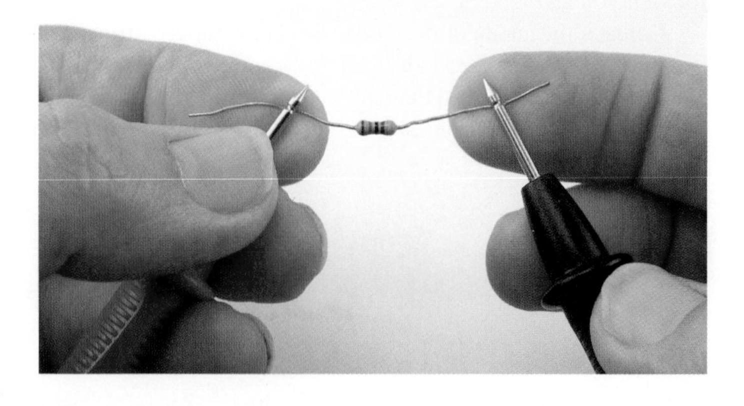

Figure 11-2 How not to measure high-value resistors

Use test leads with crocodile clips, or pin the resistor to your work surface with the flat of the test leads.

Capacitance

Some multimeters include a capacitance range. While not particularly useful for finding the value of unknown capacitors (capacitors have their value written on them), being able to test a capacitor and make sure it still has a capacitance something close to its stated value is useful.

The capacitance range on most meters is quite inaccurate, but then the values of actual capacitors—especially electrolytics—often have quite a wide tolerance.

In other words, if your meter tells you that your 100μF capacitor is actually 120μF, then that is to be expected.

Temperature

If your multimeter has a temperature range, it probably also comes with a special set of leads for measuring it, such as those shown in Figure 11-3.

The leads are actually a thermocouple that can measure the temperature of the tiny metal bead on the end of the leads. This thermometer is a lot more useful than your average digital thermometer. Check the manual for your meter, but the range of temperature is likely to be something like –40°C to 1000°C (–40°F to 1832°F).

So, you can use it to check how hot your soldering iron is getting, or if you have a component in a project that seems to be getting a bit toasty, you can use this to check just how hot it is getting.

FIGURE 11-3 Thermocouple leads for temperature measurement

AC Voltage

We have not talked about AC very much in this book. AC stands for alternating current and refers to the type of electricity you get in a home wall socket's 110V or 220V supply. Figure 11-4 shows how 110V AC household electricity voltage varies over time.

From Figure 11-4, it is apparent that the voltage actually reaches a peak of 155V and swings all the way negative to –155V. So you might be wondering why it is referred to as 110V at all.

The answer is that since a lot of the time, the voltage is quite low, at those times, it delivers very little power. So the 110V is a kind of average. It's not the normal average voltage, because that would be (110 – 110) / 2 = 0V, and because half the time it is negative.

110V is the RMS voltage (root mean squared). This is the peak positive voltage divided by the square root of 2 (1.4). You can think of this as the DC equivalent voltage. So a light bulb running on 110V AC would appear to be the same brightness as if it were running on 110V DC.

You are unlikely to need to measure AC unless you are doing something exotic and dangerous, and you should not do that unless you are very sure about what you are doing and therefore probably already knew what I just told you.

FIGURE 11-4 Alternating current

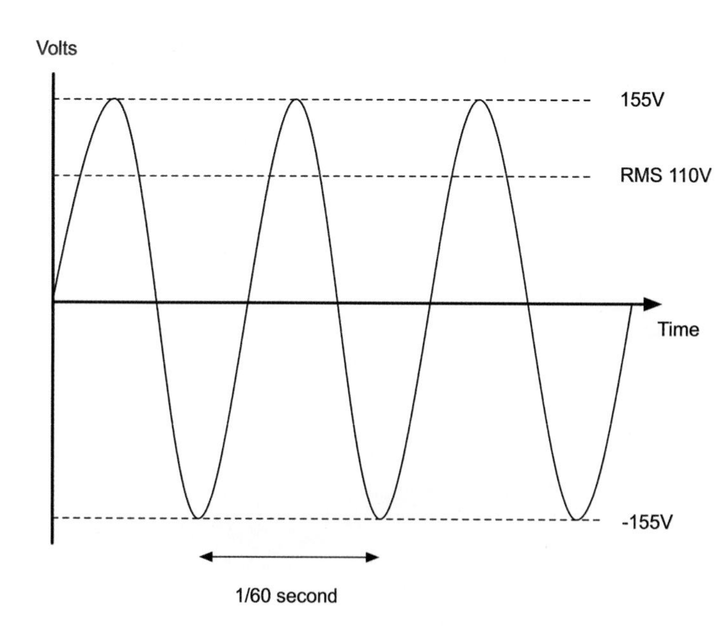

DC Voltage

We have already measured DC voltage quite a lot—mostly at the 0 to 20V range.

There is nothing much more to say about this, except to always start with the highest voltage range you believe you are about to measure and then work your way down.

DC Current

When measuring current, you will probably find that for all current ranges you will need to use different sockets on the multimeter for the positive probe lead. There is usually one connection for low currents and a separate one for the high-current ranges (20A on the author's multimeter; see Figure 11-5).

There are two important points to consider here. First, if you exceed the current range, your meter will not just give you a warning, it may well blow a fuse within the meter.

The second point is that when the probe leads are in the sockets for current measurement, there is a very low resistance between them. After all, they need to allow as much of the original current as possible to flow through them. So, if you forget that the leads are in these sockets and go to measure a voltage elsewhere in the circuit, you will effectively short-out your circuit and probably blow the fuse on your multimeter at the same time.

So, just to reiterate, if you have been using your multimeter to measure current, ALWAYS put the probe leads back to their

FIGURE 11-5 High-current measurement

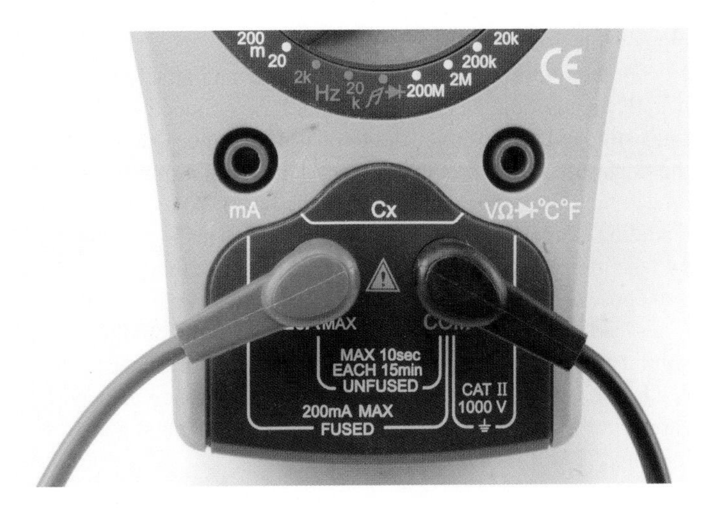

voltage sockets, as these are more likely to be used next. If you try and measure current with the leads in the voltage sockets, then all that will happen is that you get a reading of zero.

AC Current

The same argument that we gave for measuring AC voltage probably applies here. Exercise extreme caution.

Frequency

If your multimeter has a frequency setting, this can be useful. For example, in the section "How to Generate Tones with a 555 Timer" in Chapter 9 where we create an audio tone using a 555 timer, you could use this feature to measure the frequency of the tone being produced. This can be handy if you do not have access to an oscilloscope.

How to Use a Multimeter to Test a Transistor

Some multimeters actually have a transistor test socket where you can plug in a transistor. The multimeter will not only tell you if the transistor is alive or dead, but also what its gain (Hfe) is.

If your multimeter does not have such a feature, you can use the diode test feature to at least tell you if the transistor is undamaged.

Figure 11-6 shows the steps involved in testing an NPN bipolar transistor like the 2N3906.

Put the multimeter into diode test mode and attach the negative lead of the meter to the center base connection of the transistor, and the positive lead to one of the other leads of the transistor. It does not matter if it is the emitter or collector (check the transistor's pinout to find the base). You should get a reading, somewhere between 500 and 900. This is the forward voltage in mA between the base and whichever other connection you chose (Figure 11-6a). Then, move the positive lead to the other lead of the transistor (Figure 11-6b) and you should see a similar figure. If either reading is zero, either your transistor is dead, or it is a PNP type of transistor, in which case you need to carry out the same procedure but with the positive and negative leads to the multimeter reversed.

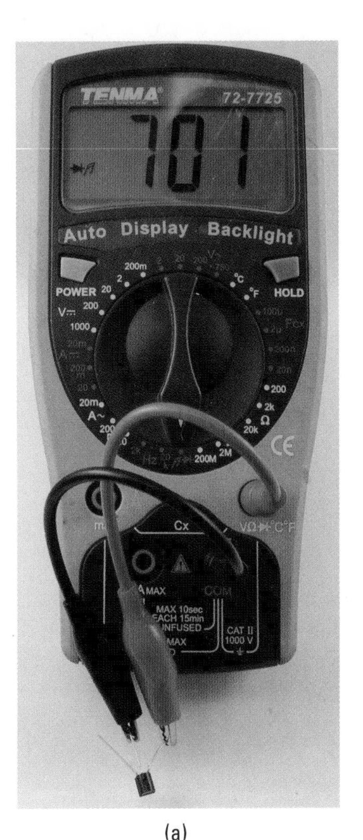

(a)

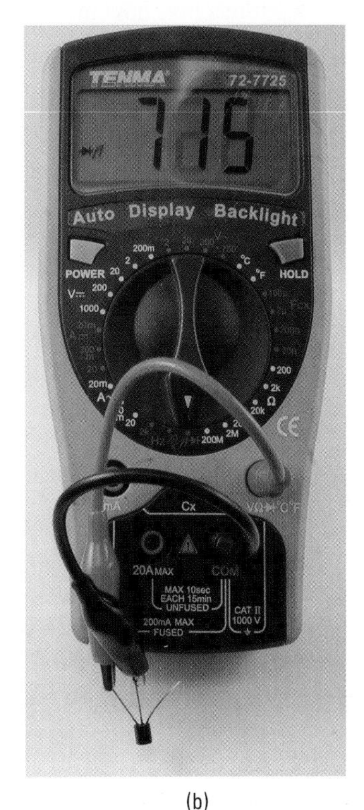

(b)

FIGURE 11-6 Testing a transistor

How to Use a Lab Power Supply

We came across a lab power supply back in Chapter 5. If you have your soldering equipment and a multimeter, then a lab power supply (Figure 11-7) is probably the next item to invest in. It will get a lot of use.

The power supply shown in Figure 11-7 is a simple-to-use basic design. In the figure, it is being used to charge a lead–acid battery. You will find that you use it to power your projects while developing them. You should be able to get something similar for under USD 100.

It plugs into your home electrical socket and can deliver up to 20V at 4A, which is more than enough for most purposes. The screen displays the voltage at the top, and the current being consumed at the bottom.

The reasons why it is more convenient than using batteries or a fixed power supply are:

- It displays how much current is being consumed.
- You can limit the current consumption.
- You can use it in constant current mode when testing LEDs.
- You can adjust the voltage easily.

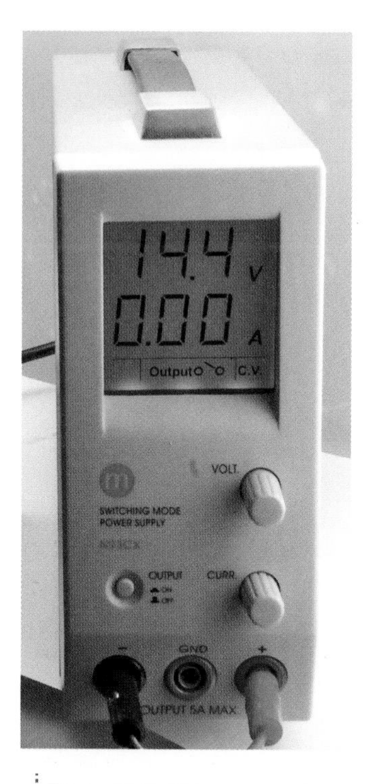

FIGURE 11-7 A lab power supply

The control panel has an Output switch that turns the output voltage on and off, and two knobs that control the voltage and current.

If I am powering up some project for the first time, I will often follow this procedure:

1. Set the current to its minimum setting.
2. Set the desired voltage.
3. Turn on the output (the voltage will probably drop).
4. Increase the current and watch the voltage rise, making sure that the current isn't rising to an unexpected level.

Introducing: The Oscilloscope

Oscilloscopes (Figure 11-8) are an indispensable tool for any kind of electronics design or test where you are looking at a signal that changes over time. They are a relatively expensive

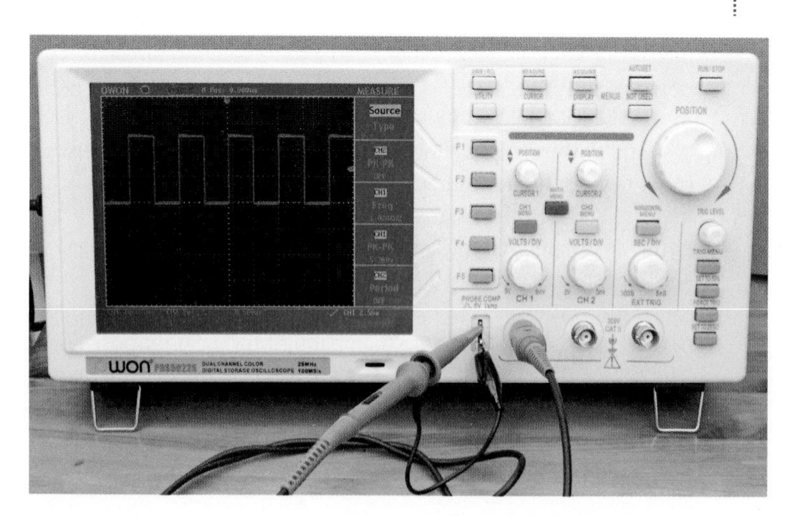

FIGURE 11-8 A low-cost digital oscilloscope

bit of equipment (from USD 200 on up) and there are various kinds. One of the most cost-effective types does not have any display at all, but connects to your computer over USB. If you don't want to risk blobs of solder on your laptop, or wait for it to boot up, then a dedicated oscilloscope is probably best.

Entire books have been written about using an oscilloscope effectively, and every oscilloscope is different, so we will just cover the basics here.

As you can see from Figure 11-8, the waveform is displayed over the top of a grid. The vertical grid is in units of some fraction of volts, which on this screen is 2V per division. So the voltage of the square wave in total is 2.5 × 2 or 5V.

The horizontal axis is the time axis, and this is calibrated in seconds. In this case, 500 microseconds (μS) per division. So the length of one complete cycle of the wave is 1000 μS—or 1 millisecond—indicating a frequency of 1 KHz.

The other advantage of an oscilloscope is that the test leads are very high impedance, which means that they have very little effect on the thing you are trying to measure.

Software Tools

As well as hardware tools for hacking electronics, there are lots of useful software tools that can help us out.

Simulation

If you like the idea of trying out electronic designs in a virtual world, you should try one of the online simulators like CircuitLab (www.circuitlab.com). This online tool (Figure 11-9) allows you to draw your circuits online and simulate how they will behave.

You will have to pick up a bit more theory than this book covers, but a tool like this can save you a lot of effort.

Fritzing

Fritzing (www.fritzing.org) is a really interesting open-source software project that lets you design projects. It is intended

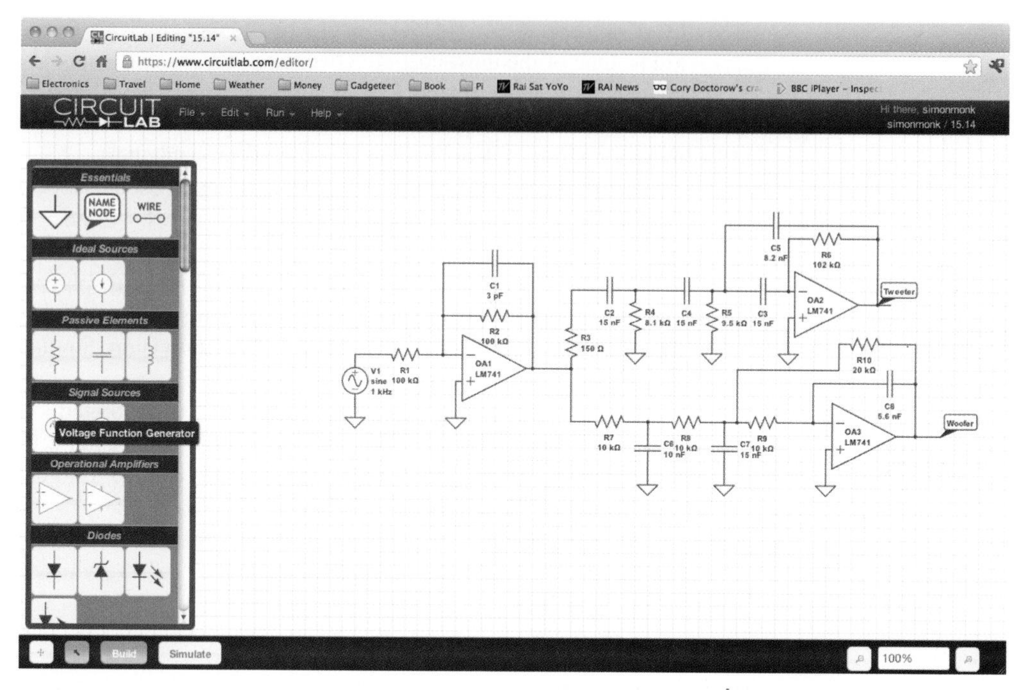

FIGURE 11-9 The CircuitLab simulator

primarily for breadboard design and includes libraries of components and modules, such as an Arduino, that can all be wired up (Figure 11-10).

EAGLE PCB

If you want to start creating your own PCBs for your electronics designs, then look for the most popular tool for this, which is called EAGLE PCB (Figure 11-11). It allows you to draw a schematic diagram and then switch to a PCB view where you can route the connections between components before creating the CAM (computer-aided manufacturing) files, which you can then send off to a PCB fabrication shop.

Creating PCBs is a subject in its own right. For more information on this, take a look at the book *Make Your Own PCBs with EAGLE: From Schematic Designs to Finished Boards* by Simon Monk (TAB, 2013).

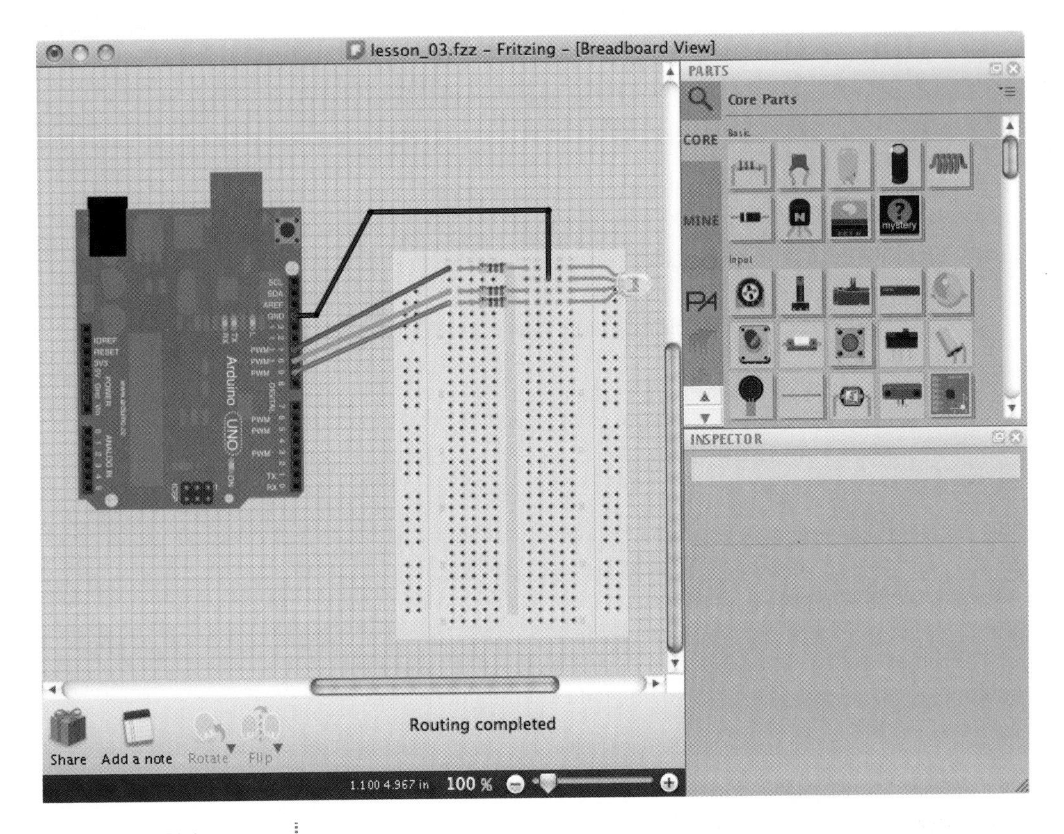

FIGURE 11-10 Fritzing

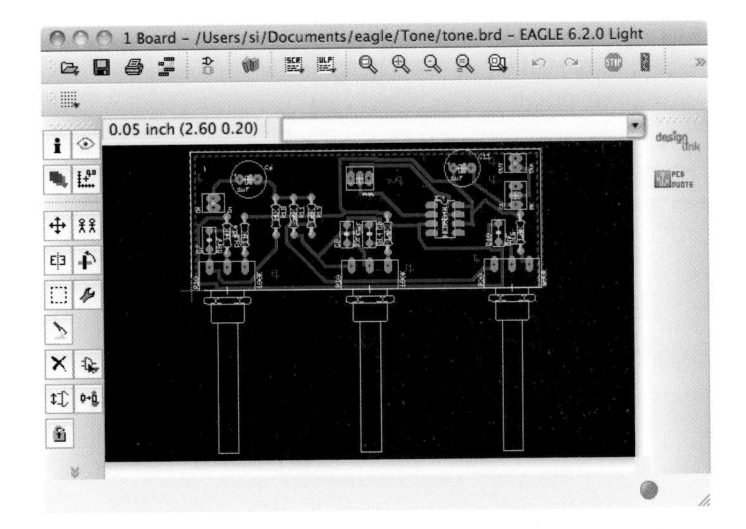

FIGURE 11-11 EAGLE PCB

Online Calculators

Online calculators can make your electronics math a whole lot easier. Some of the more useful ones are:

- **http://led.linear1.org/1led.wiz** A series resistor calculator for LEDs
- **http://led.linear1.org/led.wiz** Designed for driving large numbers of LEDs
- **www.bowdenshobbycircuits.info/555.htm** A 555 timer IC component calculator

Summary

This is the last chapter in this book and I hope it will help you get started "hacking electronics." There is much satisfaction in making something physical, or modifying a device so it does just what you want.

The line between producer and consumer is blurring more and more today as people start designing and building their own electronic devices.

The Internet offers many useful resources. The following web sites are worth a special mention:

- www.hacknmod.com
- www.instructables.com
- www.arduino.cc (for Arduino)
- www.sparkfun.com (modules and interesting components)
- www.adafruit.com (more cool stuff)
- www.dealextreme.com (bargains; search for LEDs, etc.)
- www.ebay.com (search for the same items as that in the other URLs in this list)

See also the components suppliers mentioned in the Appendix.

Appendix

Parts

Prices of components vary enormously, so please treat the following lists as a guide and shop around.

I know some people who buy almost everything on eBay. But beware. Though things are often very cheap there, occasionally they are much more expensive than at other suppliers.

I have listed part codes for the tools, modules, and so on from SparkFun and Adafruit, as these suppliers are very accessible to hobbyists and also provide good accompanying documentation. They also have distributors throughout the world, so you do not have to buy direct from either company if you live outside the U.S.

For other components, I have tried to list product codes for Mouser and DigiKey since these predominate as suppliers to hobbyists in the U.S., and also Farnell, who are UK-based but will ship to anywhere.

Please also see the book's web site (www.hackingelectronics.com) as updates for component availability will appear here.

Tools

Book Code	Description	SparkFun	Adafruit
T1	Beginner toolkit (soldering kit, pliers, snips)	TOL-09465	
T2	Multimeter	TOL-09141	
T3	PVC insulating tape	PRT-10688	
T4	Helping hands	TOL-09317	ID: 291
T5	Solderless breadboard	PRT-00112	ID: 239
T6	Solid-core jumper wire set	PRT-00124	ID: 758
T7	Red hookup wire (22 AWG)	PRT-08023	ID: 288
T8	Black hookup wire (22 AWG)	PRT-08022	ID: 290

Book Code	Description	SparkFun	Adafruit
T9	Yellow hookup wire (22 AWG)	PRT-08024	ID: 289
T10	Red multi-core wire (22 AWG)	PRT-08865	
T11	Black multi-core wire (22 AWG)	PRT-08867	
T12	Male-to-female jumper set	PRT-09385	ID: 825
T13	Desoldering braid / wick	TOL-09327	ID: 149

Components

To get yourselves a basic stock of components, you are strongly recommended to buy a starter kit of components. SparkFun sells such a set, but it doesn't include resistors, so you will need to buy a resistor set, too. Once you have these, you will own a useful collection of components that should cover the majority of what you need.

Component Starter Kits

The SparkFun Beginner Parts Kit and Resistor Kit will give you a good initial stock of parts.

Book Code	Description	SparkFun
K1	SparkFun Beginner Parts Kit (KIT-10003)	KIT-10003
K2	SparkFun Resistor Kit	COM-10969

Resistors

Book Code	Description	SparkFun	Adafruit	Other
R1	10kΩ trimpot, 0.1-inch pitch (also in K1)	COM-09806	ID: 356	DigiKey: 3362P-103LF-ND Mouser: 652-3362P-1-103LF Farnell: 9354301
R2 also in K1	LDR (also in K1)	SEN-09088	ID: 161	DigiKey: PDV-P8001-ND Farnell: 1652637
R3	500Ω trimpot			DigiKey: CT6EP501-ND Mouser: 652-3386P-1-501LF Farnell: 9355103

Capacitors

Book Code	Description	SparkFun	Other
C1	1000µF 16V electrolytic		DigiKey: P10373TB-ND Mouser: 667-ECA-1CM102 Farnell: 2113031
C2	100µF 16V electrolytic (also in K1)	COM-00096	DigiKey: P5529-ND Mouser: 647-UST1C101MDD Farnell: 8126240
C3	470nF capacitor		DigiKey: 445-8413-ND Mouser: 810-FK28X5R1E474K Farnell: 1179637
C4	100nF capacitor (also in K1)	COM-08375	DigiKey: 445-5258-ND Mouser: 810-FK18X7R1E104K Farnell: 1216438 Adafruit: 753
C5	10µF capacitor (also in K1)	COM-00523	DigiKey: P14482-ND Mouser: 667-EEA-GA1C100 Farnell: 8766894

Semiconductors

Book Code	Description	SparkFun	Adafruit	Other
S1	2N3904 (also in K1)	COM-00521	756	DigiKey: 2N3904-APTB-ND Mouser: 610-2N3904 Farnell: 9846743
S2	High-brightness white LED (5mm)	COM-00531	754	DigiKey: C513A-WSN-CV0Y0151-ND Mouser: 941-C503CWASCBADB152 Farnell: 1716696
S3	1-W Lumiled LED on heatsink	BOB-09656	518	DigiKey: 160-1751-ND Mouser: 859-LOPL-E011WA Farnell: 1106587
S4	7805 voltage regulator (also in K1)	COM-00107		DigiKey: 296-13996-5-ND Mouser: 512-KA7805ETU Farnell: 2142988
S5	1N4001 diode (also in K1)	COM-08589	755	DigiKey: 1N4001-E3/54GITR-ND Mouser: 512-1N4001 Farnell: 1651089
S6	FQP30N06	COM-10213	355	DigiKey: FQP30N06L-ND Mouser: 512-FQP30N06 Farnell: 1695498

Book Code	Description	SparkFun	Adafruit	Other
S7	LM311 comparator			DigiKey: 497-1570-5-ND Mouser: 511-LM311N Farnell: 9755942
S8	TMP36 Temperature IC	SEN-10988	165	DigiKey: TMP36GT9Z-ND Farnell: 1438760
S9	TDA7052			DigiKey: 568-1138-5-ND Mouser: 771-TDA7052AN Farnell: 526198
S10	NE555 timer IC (also in K1)	COM-09273		DigiKey: 497-1963-5-ND Mouser: 595-NE555P Farnell: 1467742
S11	Red LED 5mm	COM-09590	297	DigiKey: 751-1118-ND Mouser: 941-C503BRANCY0B0AA1 Farnell: 1249928
S12	Linear hall effect sensor			DigiKey: 620-1022-ND Mouser: 785-SS496B Farnell: 1791388

Hardware and Miscellaneous

Book Code	Description	SparkFun	Adafruit	Other
H1	4 × AA battery holder	PRT-00550	830	DigiKey: 2476K-ND Mouser: 534-2476 Farnell: 4529923
H2	Battery clip			DigiKey: BS61KIT-ND Mouser: 563-HH-3449 Farnell: 1183124
H3	Stripboard			eBay—search for "stripboard" Farnell: 1201473
H4	Pin header strip	PRT-00116	392	
H5	2A two-way screw terminal			eBay—search for "terminal block" Mouser: 538-39100-1002
H6	6V gear motor			Part of H7 eBay—search for "gear motor" or "gearmotor"
H7	Magician chassis	ROB-10825		
H8	6 × AA battery holder		248	DigiKey: BH26AASF-ND Farnell: 3829571

Book Code	Description	SparkFun	Adafruit	Other
H9	Battery clip to 2.1mm jack adapter		80	
H10	9g servo motor	ROB-09065	169	
H11	2.1mm barrel jack plug			DigiKey: CP3-1000-ND Farnell: 1737256
H12	Small solderless breadboard	PRT-09567	64	
H13	12V bipolar stepper motor	ROB-09238	324	
H14	8Ω speaker	COM-09151		
H15	Large pushbutton switch	COM-09336	559	
H16	5V relay	COM-00100		Digikey: T7CV1D-05-ND

Modules

Book Code	Description	SparkFun	Adafruit	Other
M1	12V 500mA power supply	TOL-09442	798	Note: U.S. model listed here.
M2	Arduino Uno R3	DEV-11021	50	
M3	Piezo sounder	COM-07950	160	
M4	Arduino Ethernet Shield	DEV-09026	201	
M5	PIR module	SEN-08630	189	
M6	MaxBotix LV-EZ1 rangefinder	SEN-00639	172	
M7	HC-SR04 rangefinder			eBay—search for "HC-SR04"
M8	AK-R06A RF Kit			eBay—search for "433MHZ 4 Channel RF Radio"
M9	SparkFun TB6612FNG breakout board	ROB-09457		
M10	Piezo sounder (built-in oscillator)			eBay—search for "Active Buzzer 5V"
M11	Methane sensor MQ-4	SEN-09404		
M12	Color sensing module			eBay—search for "TCS3200D Arduino"
M13	Piezo vibration sensor	SEN-09199		

Book Code	Description	SparkFun	Adafruit	Other
M14	SparkFun mic module	BOB-09964		
M15	Accelerometer module		163	Freetronics: AM3X
M16	USB LiPo charger	PRT-10161	259	
M17	Combined LiPo charger, Buck-booster	PRT-11231		
M18	Arduino LCD shield			Freetronics: LCD Keypad Shield
M19	4-digit, 7-segment display w/I2C backpack		880	
M20	RTC module		264	
M21	Arduino Leonardo	DEV-11286	849	

Index

C

H

I

J

K

L

M

N

O

P

R

S

T

U

V

W

AREF
GND
DIGITAL (PWM~)
TX
RX
ARDUINO
UNO
ON

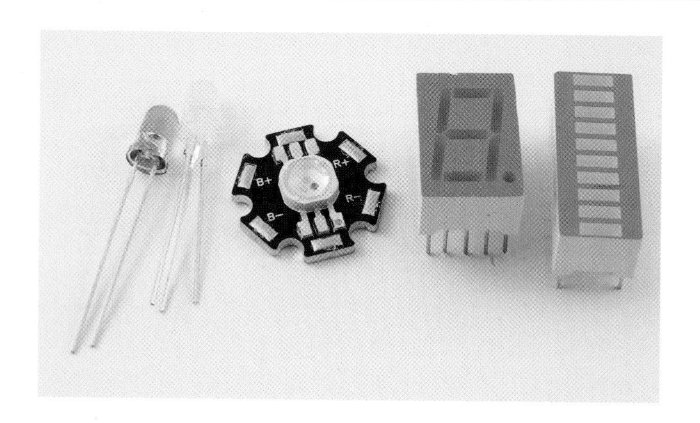
B+
R+
B–
R–

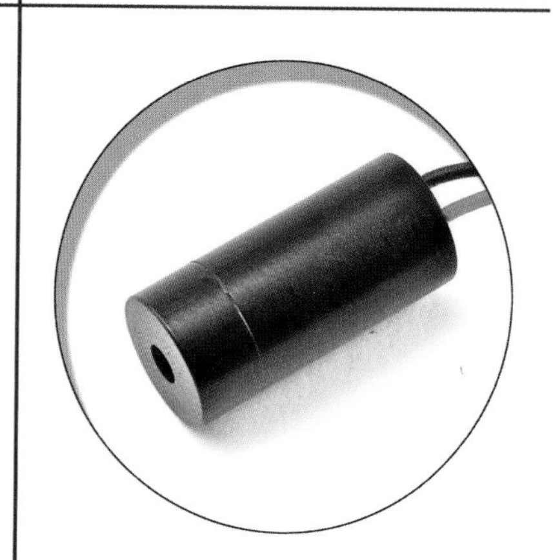

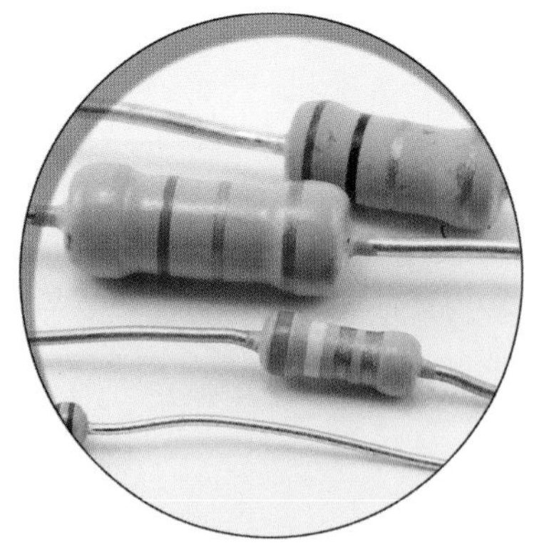

AREF
GND
13
12
~11
~10
~9
8
7
~6
~5
4
~3
DIGITAL (PWM~)
UNO
ARDUINO
TX
RX
ON
47
16V

ON
1 2 3 4

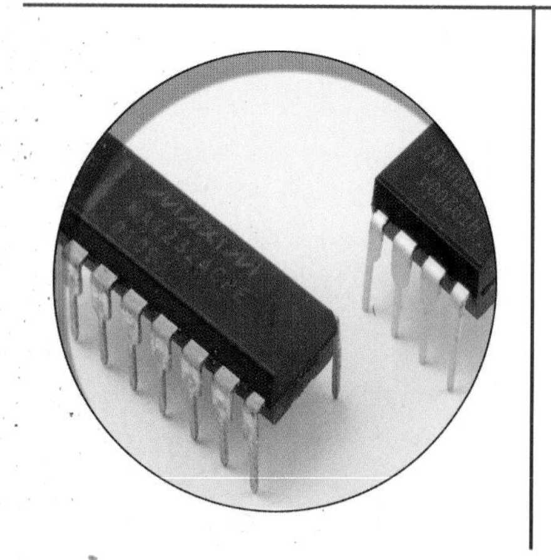

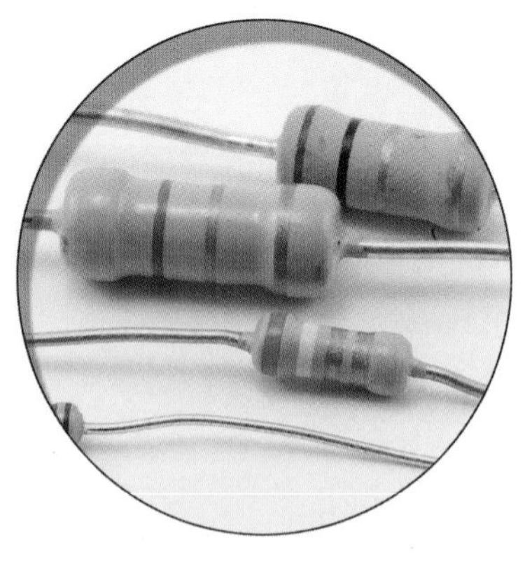